T0178997

VLSI ARCHITECTURE FOR SIGNAL, SPEECH, AND IMAGE PROCESSING

VLSI ARCHITECTURE FOR SIGNAL, SPEECH, AND IMAGE PROCESSING

Edited by
Durgesh Nandan, PhD
Basant K. Mohanty, PhD
Sanjeev Kumar, PhD
Rajeev Kumar Arya, PhD

APPLE
ACADEMIC
PRESS

First edition published 2023

Apple Academic Press Inc.
1265 Goldenrod Circle, NE,
Palm Bay, FL 32905 USA

760 Laurentian Drive, Unit 19,
Burlington, ON L7N 0A4, CANADA

CRC Press
6000 Broken Sound Parkway NW,
Suite 300, Boca Raton, FL 33487-2742 USA

4 Park Square, Milton Park,
Abingdon, Oxon, OX14 4RN UK

© 2023 by Apple Academic Press, Inc.

Apple Academic Press exclusively co-publishes with CRC Press, an imprint of Taylor & Francis Group, LLC

Library and Archives Canada Cataloguing in Publication

Title: VLSI architecture for signal, speech, and image processing / edited by Durgesh Nandan, PhD, Basant K. Mohanty, PhD, Sanjeev Kumar, PhD, Rajeev Kumar Arya, PhD.
Names: Nandan, Durgesh, editor. | Mohanty, Basant K., editor. | Kumar, Sanjeev (Research mentor), editor. | Arya, Rajeev, 1985- editor.
Description: First edition. | Includes bibliographical references and index.
Identifiers: Canadiana (print) 20220200424 | Canadiana (ebook) 20220200572 | ISBN 9781774637302 (hardcover) | ISBN 9781774637319 (softcover) | ISBN 9781003277538 (ebook)
Subjects: LCSH: Integrated circuits—Very large scale integration.
Classification: LCC TK7874.75 .V57 2023 | DDC 621.39/5—dc23

Library of Congress Cataloging-in-Publication Data

CIP data on file with US Library of Congress

ISBN: 978-1-77463-730-2 (hbk)
ISBN: 978-1-77463-731-9 (pbk)
ISBN: 978-1-00327-753-8 (ebk)

About the Editors

Durgesh Nandan, PhD
Research Mentor and Account Manager,
Accendere Knowledge Management Services Pvt.
Ltd. (a subsidiary of CL Educate Ltd)

Durgesh Nandan, PhD, is a Research Mentor and Account Manager at Accendere Knowledge Management Services Pvt. Ltd. (100 % subsidiary of CL Educate Ltd). He formerly served as Assistant Professor and Head of Department in the Department of Electronics and Communication Engineering, IASSCOM Fortune Institutes of Technology, India, as well as a guest faculty under the Special Manpower Development Program for Chips to System Design (SMDP-C2SD) in the Department of Electronics & Communication Engineering, NIT, Patna, India. He is session chair, technical program committee chair, reviewer, and member of more than 50 national and international conferences. He is the author or a co-author of more than 90 papers. He is inventor or co-inventor of a published Indian patent and six Indian patents in process of filing. He is the author or a co-author of two books. His research interests include computer arithmetic, VLSI architecture for signal processing applications, speech processing, hardware architecture of real time big data/AI applications, and Internet of Things. He founded the prestigious JSS Fellowship for 2014 to 2018 by Jay-Prakash Sewa Sasthan. He also received a "Young Personality of the Year Award (under 40 years)" in 2019 by the International Academic and Research Excellence Awards (IARE–2019). He also awarded at "I2OR Preeminent Researcher Award 2019" in 2019 for remarkable contribution in the field of VLSI and DSP by the International Institute of Organized Research.

Dr. Nandan earned a PhD from the Department of Electronics & Communication Engineering, Jaypee University of Engineering and Technology, Guna, Madhya Pradesh, India, with the specialization in VLSI. His MTech, with specialization in Microelectronics & VLSI Design, and his BS (Engineering) were awarded by E.C.E from Rajeev Gandhi Technical University, Madhya Pradesh, India.

Basant K. Mohanty, PhD

*Professor and Associate Dean (Research),
Mukesh Patel School of Technology,
Management and Engineering, Narsee
Moonjee Institute of Management Studies,
Maharashtra, India*

Basant Kumar Mohanty, PhD, is a Professor and Associate Dean (Research) at the Mukesh Patel School of Technology, Management and Engineering, Shirpur Campus, at the Narsee Moonjee Institute of Management Studies (Deemed-to be University), Mumbai, Maharashtra, India. He has 26 years of teaching and research experience in the field of science and engineering from various universities within India and abroad. He formerly has worked with various institutions during his career, starting his professional teaching career at Odisha Education Service, Class-II Gazetted, and he joined SKCG (Autonomous) College, Paralakhemundi, Odisha. Later he moved on to join BITS Pilani- Rajasthan, Mody University of Science and Engineering-Rajasthan, Jaypee University of Engineering and Technology, Guna-Madhya Pradesh, India. He has also been associated with Qatar University and the School of Computer Engineering, Nanyang Technological University, Singapore, for postdoctoral research. He is an Associate Editor for the Journal of Circuits, Systems, and Signal Processing, a senior member of IEEE Circuits and System Society, and a life-time member of the Institution of Electronics and Telecommunication Engineering, New Delhi, India. His research interests include design and implementation of reconfigurable architectures for domain-specific signal processing applications, approximate computation, stochastic computation, and compressed sensing. He has contributed more than 60 technical papers to various reputed journals and conference proceedings, which includes over 34 SCI journal publications with over 17 IEEE Transaction papers. He has research collaboration with faculties of top technical universities within and outside India. He is a senior member of IEEE Circuits and System Society and lifetime member of the Institution of Electronics and Telecommunication Engineering, New Delhi, India.

Dr. Mohanty received his undergraduate and postgraduate degrees in Physics with a specialization in electronics and his PhD in VLSI for digital signal processing (DSP).

Sanjeev Kumar, PhD
Research Mentor, CL Educate Ltd., India

Sanjeev Kumar, PhD, is working as a Research Mentor in CL Educate Ltd., India. He served as Assistant Professor in the Department of Electronics and Communication Engineering, TIT Group of Institutes, India, as well as Assistant Professor in the Electronics and Communication Engineering Department at the Oriental Group of Institutes, Bhopal, India. He had a six-month research experience as a project fellow at the Division of MMIC in the Defence Research and Development Organisation, Delhi, India. He is technical program chair, reviewer, and member of more than 40 refereed national/international conferences. He is the author or a co-author of more than 65 papers, which were published in SCI, Scopus, and peer-reviewed international journals and conference proceedings. His research interests are in design and modelling of microstrip antennas, digital signal processing, metamaterials, and RF circuit design. He founded the prestigious "JSS Fellowship" for 2014 to 2018 at Jayaprakash Sewa Sasthan, Jaypee Institute of Information Technology, Noida, Uttar Pradesh, India.

Dr. Kumar earned his PhD in design and analysis of UWB and MIMO antenna systems for communication systems from Jaypee University of Engineering and Technology, Guna, India. He did his MTech degree with specialization in microwave-electronics at the University of Delhi, India, and his Bachelor of Engineering degree in electronics and communication engineering from Rajiv Gandhi Technical University, Bhopal, India.

Rajeev Kumar Arya, PhD
Assistant Professor, Department of Electronics and Communication Engineering, National Institute of Technology, Patna, Bihar, India

Rajeev Kumar Arya, PhD, is currently an Assistant Professor with the Department of Electronics and Communication Engineering at the National Institute of Technology, Patna, India. Dr. Rajeev Kumar Arya received the Engineering Degree in Electronics and Communication Engineering from Government Engineering College, Ujjain, (RGPV University, Bhopal) 2008, India, and the Master of Technology in Electronics and Communication Engineering from Indian Institute of Technology (ISM), 2012, Dhanbad, India. He received a PhD degree in Communication Engineering from the Indian Institute of Technology (IIT Roorkee) 2016, Roorkee, India. He has received the Ministry of Human Resource Development Scholarship (MHRD India) during MTech and PhD. He has been working as an Associate Professor in the Department of Electronics and communication engineering at CMR Engineering College Hyderabad. His current research interests are in wireless communication, soft computing techniques, cognitive radio, signal processing, communication systems, and circuits design. He has published many articles in international journals, conferences, and received the best paper award at ICCET-2019. He serves as Guest Editor International Journal of Information Technology and Web Engineering (IJITWE) IGI Global Publishers USA, International Journal of Computational Systems Engineering INDERSCIENCE Publishers-UK and Reviewer for IEEE Communication letter, some journals of Springer Publisher, SCI/SEC-E. Scopus indexed journals and some conferences. He is a member of the IEEE, ISRD, and the IAENG. He is an active reviewer in many reputed international journals.

Contents

Contributors

Rallabhandy Abhinaya
Department of Electronics and Communication Engineering, Institute of Aeronautical Engineering, Dundigal – 500043, Hyderabad, Telangana, India

Narayanam Balaji
Department of Electronics and Communication Engineering, JNTUK University College of Engineering, Kakinada, Andhra Pradesh – 533003, India

Naveen Kumar Challa
Department of Electronics and Communications Engineering, Vignan's Foundation for Science, Technology, and Research (Deemed to be University), Guntur, Andhra Pradesh – 522213, India

Abhishek Choubey
Department of Electronics and Communication Engineering, Sreenidhi Institute of Science and Technology, Hyderabad – 501301, Telangana, India

Shruti Bhargava Choubey
Department of Electronics and Communication Engineering, Sreenidhi Institute of Science and Technology, Hyderabad – 501301, Telangana, India

Magnanil Goswami
Accendere Knowledge Management Services Pvt. Ltd., New Delhi – 110044, India,
E-mail: magnanil.goswami@gmail.com

Birinderjit Singh Kalyan
I. K. Gujral Punjab Technical University, University in Kapurthala, Punjab, India

Harpreet Kaur
Department of Electronics and Communication Engineering, Guru Nanak Institute of Technical Campus Hyderabad, Telangana, India, E-mail: kaurharpr@gmail.com

A. Kavitha
Veltech Multitech Dr. Rangarajan and Dr. Sankuntala Engineering College, Chennai, Tamil Nadu, India

C. S. N. Koushik
Department of Electronics and Communication Engineering, Sreenidhi Institute of Science and Technology, Hyderabad – 501301, Telangana, India, E-mail: koushikcsn@gmail.com

Roopa R. Kulkarni
Department of Electronics and Communication Engineering, Dayananda Sagar Academy of Technology and Management, Bengaluru, Karnataka, India, E-mail: roopa.patavardhan@gmail.com

S. Y. Kulkarni
REVA Institute of Technology and Management, Bengaluru, Karnataka, India

Gundugonti Kishore Kumar
Department of Electronics and Communication Engineering, V. R. Siddhartha Engineering College, Vijayawada, Andhra Pradesh – 520007, India, E-mail: kishore.chiya@gmail.com

Kurra Anil Kumar
Department of Electronics and Communications Engineering, Vignan's Foundation for Science,
Technology, and Research (Deemed to be University), Guntur, Andhra Pradesh, India,
E-mail: kakumar94@gmail.com

N. Ashok Kumar
Department of Electronics and Communication Engineering, Sree Vidyanikethan Engineering College,
Tirupati, Andhra Pradesh, India, E-mail: ashoknoc@gmail.com

Putta Manoja
Department of Electronics and Communication Engineering, Institute of Aeronautical Engineering,
Dundigal – 500043, Hyderabad, Telangana, India

Madhusmita Mohanty
Department of Electrical and Electronics Engineering, National Institute of Technology, Puducherry,
Karaikal, India

Durgesh Nandan
Aditya Engineering College, Surampalem, Andhra Pradesh, India

D. Naresh
Department of Electronics and Communication Engineering, Sreenidhi Institute of Science and
Technology, Hyderabad – 501301, Telangana, India

Usha Rani Nelakuditi
Department of Electronics and Communications Engineering, Vignan's Foundation for Science,
Technology, and Research (Deemed to be University), Guntur, Andhra Pradesh – 522213, India

Nakka Nikhil
Department of Electronics and Communication Engineering, Institute of Aeronautical Engineering,
Dundigal – 500043, Hyderabad, Telangana, India

Khushboo Pachori
Department of Electronics and Communication Engineering, Guru Nanak Institute of Technical Campus
Hyderabad, Telangana, India

Subhransu Padhee
Department of Electrical and Electronic Engineering, Aditya Engineering College, Surampalem,
Andhra Pradesh, India, E-mail: subhransupadhee@gmail.com

Ambarish Panda
Department of Electrical and Electronics Engineering, Sambalpur University Institute of Information
Technology, Burla, Odisha, India

S. Vishnu Priyan
Kingston Engineering College, Vellore, Tamil Nadu, India

Merugu Rachana
Department of Electronics and Communication Engineering, Institute of Aeronautical Engineering,
Dundigal – 500043, Hyderabad, Telangana, India

Shaik Sadulla
Department of Electronics and Communication Engineering, KKR, and KSR Institute of Technology
and Sciences, Vinjanampadu, Andhra Pradesh – 522017, India

Pittala Chandra Shekar
Department of Electronics and Communication Engineering, MLR Institute of Technology,
Dundigal – 500043, Hyderabad, Telangana, India

Balwinder Singh
Center for Development of Advanced Computing (C-DAC), Mohali, Punjab, India

P. Venkatramana
Sree Vidyanikethan Engineering College, Tirupati, Andhra Pradesh, India

Vallabhuni Vijay
Department of Electronics and Communication Engineering, Institute of Aeronautical Engineering, Dundigal – 500043, Hyderabad, Telangana, India, E-mail: v.vijay@iare.ac.in

Abbreviations

ADC	analog to digital converter
ADP	area-delay product
AHDLs	analog hardware description languages
ANN	artificial neural network
APB	advanced peripheral bus
ARM	advanced RISC machine
BCV	between-class variance
BE	booth encoder
BEU	booth encoder unit
BPNN	backpropagation NN
BS	booth selector
CAD	computer-aided design
CBWFQ	class-based weighted fair queuing
CCA	connected component analysis
CCM	continuous conduction mode
CFN	capacitive feedback network
CHW	convolutional heterogeneous window
CMOS	complementary metal-oxide-semiconductor
CNOC	clos NOC
COS	course of service
CPD	critical path delay
CPLD	complex programmable logic device
CPU	central processing unit
CRP	challenge-response pair
DCT	discrete cosine transform
DMA	direct memory access
DPWM	digital pulse width modulation
DSP	digital signal processing
DVFS	dynamic voltage and frequency scaling
DVS	dynamic voltage scaling
DWT	discrete wavelet transform
EADP	excess-area-delay-product
EMI	electromagnetic interference
EPDP	excess-power-delay-product

EX	execute
FET	field-effect transistor
FF	fast-fast
FFNN	feed-forward NN
FFT	fast Fourier transform
FIR	finite impulse response
FLANN	fast-learning artificial neural network
FPGA	field-programmable gate array
FS	fast-slow
FSM	finite state machine
GDI	gate diffusion input
GP	geometric programs
HCI	hot carrier injection
HDintra	intra-chip hamming distance
HH	high-high
HL	high-low
HOPE	hotspot prevention
HTA	hardware tree-based multicast routing algorithm
IC	integrated circuit
ID	instruction decode
IF	instruction fetch
IIR	infinite impulse response
IM	instruction memory
IoT	internet of things
IP	intellectual property
ITRS	International technology roadmap for semiconductors
I-Type	immediate type
LANCELOT	large and nonlinear constrained extended Lagrangian optimization technique
LFSR	linear feedback shift register
LH	low-high
LL	low-low
LLA	level line angle
LNAs	low-noise amplifiers
LRM	language reference manual
LSB	least significant bit
LSD	line segment detector
LSI	large-scale integration
LT	hyperlink traversal

LUT	lookup table
MAC	multiply and accumulate
MBFF	multi-bit FF
MC	Monte Carlo
MCM	multiple constant multiplications
MEM	memory access
ML	machine learning
MLFF	multilayer feed-forward
MLP	multilayer perceptron
MP	multi-path
MPSoC	multi-processor system-on-chip
MSI	medium-scale integration
MTCMOS	multi-threshold CMOS
NBTI	negative bias temperature instability
NCH	normalized cumulative histogram
NCIA	normalized cumulative intensity area
NoC	network-on-chip
PBTI	positive bias temperature instability
PCNN	pulse coupled neural network
PDE	partial differential equation
PDP	power-delay-product
PE	processing element
PIT	progressive image transmission
PPA	partial product array
PQs	priority queues
PST	power state table
PU	processing unit
PUFs	physical unclonable functions
PWM	pulse width modulation
QCA	quantum cellular automata
QoS	quality of service
RA	recursive architecture
RAG	region adjacency graph
RB	row buffer
RC	routing calculation
RTL	register transfer level
SA	simulated annealing
SA	swap allocator
SF	slow-fast

SoC	system on chip
SOI	silicon-on-insulator
SOM	self-organizing map
Src1	source-1
Src2	source-2
SS	slow-slow
SSI	small-scale integration
SST	secure split test
SSTA	statistical static timing analysis
ST	switch traversal
S-XY	surrounding XY
TCAD	technology computer-aided design
TDDB	time-dependent dielectric breakdown
TPA	test pixel address
TSMC	Taiwan Semiconductor Manufacturing Company
TT	typical-typical
UPF	unified power format
UPWM	uniformly sampled PWM
VGS	voltage across the gate to source
VHDL	VHSIC hardware description language
VLSI	very large-scale integration
Vth	threshold voltage
WB	write back
WRED	weighted discard that is random early technology

Preface

Knowledge of very-large-scale integration (VLSI) is one of the basic needs to understand the hardware prospect of engineering. Day-by-day, its growth explores new areas of research. It plays a most important role in the performance of digital systems, digital signal processing (DSP), image processing applications, hardware security, quantum computing, etc. It is closely related to data representation schemes. Especially for DSP systems, its algorithms have a number of interesting characteristics, which can be exploited in the design of the arithmetic circuits, so that they can be implemented more efficiently in terms of "computation time, chip area, and power consumption." Nowadays, many handheld portable battery-operated devices that require efficient, errorless, and low-power arithmetic operation.

This book presents recent useful research in the fields of computer arithmetic and explores the benefit of various arithmetic circuits, their digital implementation schemes, performance considerations, and utility of computer arithmetic in various new applications. It covers how computer arithmetic is contributing to the era of quantum computing, hardware security, image processing, biomedical equipment, artificial intelligence, neural networks, and stochastic computing.

This edited volume was prepared by leading researchers in the field of computer arithmetic and computer arithmetic architecture based on different domain applications. Various new research of arithmetic circuits is presented on relevant state of the art, possible applications, and scope for future research.

Chapter 1 discusses the importance of wavelet transform to represent the image as sub-integratable square functions, which can be frequently utilized in the medical and automotive industries or in any sector where the images are used. This chapter covers various types of the wavelet transform, where each type can be a single or a multi-level designed architecture. This chapter covers various schemes like lifting, flipping, convolution, etc., to analyze the outcomes at a certain throughput rate. It covers how design filters with sharp cut-off frequencies play a major role in removing unnecessary elements from pictures.

Chapter 2 addresses the efficiency of multipliers that has a significant effect on device functionality, particularly in the processing of signals and images. The author of the "Sum of Power-of-two (SOPOT)" multiplier is discussed in this chapter. In the SOPOT architecture, unnecessary shifters and multiplexers are omitted, the new "Canonical-Sign Digit (CSD)" representation is proposed, and the register dimension is reduced.

Chapter 3 presents a 16-bit Booth multiplier with a modified 2D-DWT architecture. In this work, there is no need to generate all the partial products; only the necessary products required by the coefficient are sufficient. It is observed that for performing the execution, the proposed architecture takes a smaller number of clock cycles than the existing architecture so that it increases the speed. For the need of comparison with effective results, the proposed method, CSD, and Booth multiplier, the synthesis results are done using Verilog HDL in Cadence Genus 90 nm technology.

Chapter 4 addresses the small size MOSFET operational problems such as increasing gate-oxide leakage, enhanced junction leakage, high sub-threshold conduction, and reduced output resistance. In Chapter 4, the authors discuss that FinFET would conquer the challenges described above. Taking into account the appealing characteristics of FinFET, an app is an ALU. The arithmetic and logical operations in the digital processor are performed via the Arithmetic Logic Unit (ALU). ALU 8-bit power efficient is built with the Full Adder (FA) and gate diffusion input (GDI), which has been chosen by the designer for the realization of digital combination circuits at minimal power consumption.

Chapter 5 discusses the semiconductor supply chain facing several vulnerabilities, such as counterfeiting, cloning, tampering, etc. Hence to prevent semiconductor counterfeits from adversaries, physically unclonable functions (PUFs) have emerged to be one of the lightweight security primitives in hardware cryptography and offer robust security against attacks. The authors suggest a multiplexer-decoder-based arbiter PUF using CMOS 45 nm technology and simulated using cadence virtuoso. The obtained results estimated security metrics like uniqueness, reliability, and randomness.

Chapter 6 discusses that PUFs should be robust against temporal variations in circuits, and the effect of temporal changes in circuits has influenced the reliability. Out of the several factors, aging is one of the major and critical factors that flip the PUF response. In practice, it is essential to identify the major factors responsible for aging, such as bias temperature instability (BTI), hot carrier injection (HCI), time-dependent dielectric breakdown (TDDB), electro-migration, etc. This chapter mainly addressed the impact of aging on strong PUFs and estimated the aging phenomenon on MUX-decoder PUF architecture by applying statistical analysis using cadence virtuoso 45 nm CMOS technology.

Chapter 7 discusses various power management design techniques and the specification of power state tables (PSTs), switching networks, signal isolation, state retention, and restoration of APB protocol. The peripherals like UART, Keypad, Timer, and other peripheral devices are connected to the bus architecture using low bandwidth and low-performance buses.

Chapter 8 discusses network-on-chip (NoC), which has been coming out as a strong aspect that decides the functioning and power utilization of several principal processes. The author aims to optimizes the route assigning time to effectively transmitting the data from the source to destination in this chapter. The proposed system creates a mesh topology depending on the router structure pattern and minimizes the route-sharing methods using a hybrid system.

Chapter 9 is based on the three important things to design NoC is topology: routing mechanism, and switching algorithm. The routing algorithm and network topology are the two important aspects of on-chip communication in NoC. This chapter is dedicated to the routing strategy for NoC architecture.

Chapter 10 discusses a self-driven clock gating technique that uses the XNOR gate as the comparator. The process compares the current input to the previous output, and then, to obtain the gated output, the output is eventually ANDed. This technique can be applied to any high-speed systems such as DSP processors, image processor, bio-medical processors, and in general to any processor that requires power optimization. The authors of this chapter have applied this technique to an academic state-of-art design of a 32-bit RISC processors, achieving a total power reduction of nearly 16% to 18%, with an increase in the area of 3% to 5% for a different set of test cases, using 45 nm process technology. This can be further integrated into an electronic design automation commercial backend design flow, wherein the algorithm of the tool can select the proposed clock gating techniques as the integrated clock gating cell.

Chapter 11 presents an analysis that attempts to rephrase these predicaments by means of a semi-empirical approach based on device-circuit integration. This amalgamated therapy utilizes mathematical modeling to exploit the device-circuit fundamentals and offers advantages in terms of cost-minimization, promptness, and reliability in obtaining solutions for (almost) any real-time IC design problem with industry acceptable extent of fidelity.

Chapter 12 analyzes the QCA based flip-flops and proposed novel layouts of the shift register and ring counter with less QCA cells and better performance parameters. The SISO shift register and ring counter are structured using a D flip flop, which is redesigned using 38 cells that show 42% less complexity than previous structures.

Chapter 13 provides a systematic overview of the design and performance of voltage-mode digital control of a DC-DC converter. Different design aspects of digital control have been highlighted, and digital implementation of PID controller and pulse width modulator scheme have been discussed. A case study of the DC-DC buck converter is used to illustrate the working of the digital controller.

Chapter 14 discusses how image processing has evolved at a fast pace due to its numerous applications in the field of medical, military, satellite imaging, and so forth. Image segmentation is a crucial element of image processing. It is necessary for the study of a specific object in an image by highlighting its boundary. Several algorithms and techniques have been developed with time to solve the problem of image segmentation. However, for its real-time implementation, the algorithm needs to be realized on hardware. The parallel processing capability and reconfigurability of FPGA make it an ideal option for real-time implementation. In this chapter, the hardware prospects of various image segmentation techniques are reviewed and discussed.

The primary readership for this edited volume will be senior undergraduate and graduate students in computer science, computer engineering, and electrical and electronics engineering. It may be used as a textbook for courses on VLSI, as well as a reference book. The prerequisites required for a complete understanding of the book include fundamentals of digital logic and basic ability in mathematics.

CHAPTER 1

Evolution of 1-D, 2-D, and 3-D Lifting Discrete Wavelet Transform VLSI Architecture

C. S. N. KOUSHIK, ABHISHEK CHOUBEY, SHRUTI BHARGAVA CHOUBEY, and D. NARESH

Department of Electronics and Communication Engineering, Sreenidhi Institute of Science and Technology, Hyderabad – 501301, Telangana, India, E-mail: koushikcsn@gmail.com (C. S. N. Koushik)

ABSTRACT

Wavelet is generally used to represent the image as sub-integratable square functions that are orthogonal to each other and used to reduce the effects of noise upon processing from the image. It plays an important role to compress the image and analyze it accordingly to generate the outputs. It can be used in the fields like medical and automobiles or in any field which utilizes pictures as its input. With the support of the neural networks, the efficiency of the system can be increased to a greater extent and generate the outputs in an intelligent way and in a better way selecting the best option from the various alternatives. The architecture can be of various types like 1-D, 2-D, 3-D, etc., where each can be a single or a multilevel designed architecture. There is a huge requirement of the MAC units along with the memory elements and a pipelining architecture to enhance the computational speed. It can be processed by the schemes like lifting, flipping, convolution, etc., to analyze the outcomes at a certain throughput rate. It is achieved by designing filters with sharp cut-off frequencies that play a major role in removing unnecessary elements from the picture.

1.1 INTRODUCTION

A wavelet, in general, can be considered as oscillations or vibrations that are sinusoidal in nature but with certain amplitude and frequency but in domains like signal processing. It is considered a mathematical function, which is useful in digital signal processing (DSP) and image compression techniques. It is mainly used in the field of image processing or DSP in anyone of the sectors like medical, researching, development of a convolution neural network, mobile communication, social media, and even in the field of defense for image analysis. With the use or the help of the VLSI, one can develop an integrated circuit (IC) chip or a module that can be used for other major tasks that implement image processing. The design and the architecture can be enhanced by considering the various factors like the number of rows and columns that is the dimensionality that one needs in order to achieve the desired task outcome and the adder design or the multiplier design and the memory storage requirements for the dimensionality that has been chosen by the developer. The IC can be designed either in a semi-custom manner or in a full custom manner based on the designer's choice. The semi-custom model can be done with the help of the hardware description languages, whereas the full custom design can be done with the help of the transistor level design with the end-to-end connections that is designing the entire architecture in an electrical circuit. The entire design needs to be fabricated in order to provide safety to the design, but the cost of the fabrication is quite high as usual. The area-delay-product (ADP) can be found to estimate the optimization of the entire design and the efficiency of the entire design in terms of the delay and speed criteria [1–5, 14–25].

Discrete wavelet transform (DWT) is basically used to transform the wavelets into the frequency domain that are discretely sampled in the frequency domain. In general, any wavelet transform can be represented as in an equation manner:

$$\varphi(y) = \sum_{m=-\infty}^{\infty} A_m \varphi(s_k - m) \tag{1.1}$$

Whereas a Fourier transform equation is given as, continuous frequency domain:

$$F(j\omega) = F(X) = \int_{-\infty}^{\infty} f(t) e^{-j\omega t} dt \tag{1.2}$$

The DWT is quite similar to that of the Fourier transform, but it differs with it in the aspects like the truncation, but it can be still considered very much similar to that of the Fourier transform. In the frequency domain, there is a chance to get ringing effects in the main signals that may tend to interfere with the signal and have its own impact based on its amplitude but whereas in the case of the DWT, there are no chances of the ringing and interference among the signals and hence have optimal results especially at the time of the filtering, as quite a large number of the filter banks are used in order to filter the required frequency components from the signal which when sampled and digitized at or above the Nyquist sampling rate. The ringing effects are seen to take the sigmoid distribution, which has damped oscillations outside the major intensity level of the signal. The DWT signals can be taken as a finite duration signal of small waves and limited amplitude [1, 2, 17, 30–55].

The main difference between the DCT (discrete cosine transform) and DWT is regarding the loss or error handling capability in the images that are inputted, and the sampling was done to the signal in the corresponding equivalent domain. They both are compared in the frequency domain, and the major difference can be highlighted by the sampling of the signal in the DWT, whereas the DCT ed signal is not sampled in the frequency domain, and hence the signal is considered to be continuous. As the signal is considered to be continuous in nature, the probability of the noise or the error to creep into the signal increases even though the signal to noise ratio is quite high, but there is still a probability of noise to creep into the signal yet [1, 2, 17, 30–55].

The comparison can be depicted easily by the use of the filter banks that are used to separate out or filter out the signals that are required and then give the required part to the operation in the frequency domain. The efficiency of the filters or the sharpness of the filters determines the proximity of the signal to become as ideal as possible or high accurate as possible and to have the least effect on the noise signal. The noise or the errors that are generated in the images at the time of the sampling of the signal or at the time of the computations or processing of the sampled input, needs to be handled with utmost care in order to generate the exact replica or desired output of the task that one needs. The filters used are either finite (FIR) or infinite (IIR) impulse response filters that must have their frequency response as sharp as possible in order to filter the signal components or wavelets and even the components of the noise. An array of the filter banks comprising the low pass filters and the high pass filters, respectively, are used to analyze the signal. The filter banks can be cascaded, and the required components can be filtered. The stability of the filters depends on the location of the poles of the transfer function in the s-plane or in the z-plane. The sampling rate of the signal can either be interpolated or decimated as per one's requirement in the task to achieve the outcome. In order to reconstruct the signal, it must be made to pass through the low pass filter or a bandpass filter based on the requirement of the desired frequency component, which can be a Kaiser or Chebyshev or a butterwort approximation filter that satisfies the requirement or the sharpness of the filter that one desires in operation for the accuracy. The frequency response of the filter must have been having a proper cut-off frequency or centered on the resonant frequency [1, 2, 14–25, 30–55].

There are various types of wavelet transformation techniques that have their own unique application, based on the requirement one needs. The techniques are:

- Continuous wavelet;
- Discrete wavelet;
- Fast wavelet;
- Multi-resolution analysis;
- Haar wavelet, etc.

These techniques are used when there is a need or a requirement to reduce the size of the image or, more specifically, reduce or compress the image by smoothening it and have an efficient and ease of data transfer and processing. The compression can be lossy and lossless based on the accuracy of the entire operation that one desires. It plays a major role in image processing by analyzing the data at different frequency components or approximations based on the type of filters used at the time of the processing.

In signal processing, the wavelets remove the effects of the noise and are used to compress the image by the methods like the JPEG method or to convert it into a PNG file, which is a portable network graphic file. Hence the minute errors or the noise that creeps into the signal can be eliminated by the signal to a large extent. As the image is subdivided into the sub-bands of low-low (LL), low-high (LH), high-low (HL), and high-high (HH) over the entire original spectrum of the signal, a wide range of the filters are required in order to segregate them and to do the processing (Figure 1.1) [2].

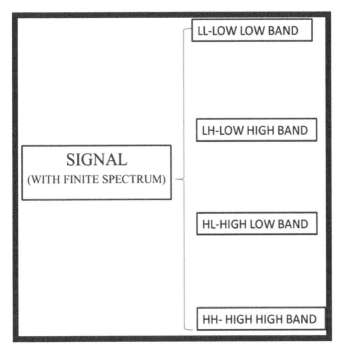

FIGURE 1.1   Signal decomposition.

1.2   TYPES OF WAVELETS

There are various representations of the wavelets based on the dimensionality of the data on uses and for the sake of the computations as well. The computations can be carried out in the form of the array or in a matrix form. Based on the matrix computations, one can determine the bit handling capacity and the speed and delay of the entire operations governing the entire efficiency of the task and even consider the factors like area and the power parameters of an IC design. Initially, the image is then sampled at a sampling frequency above the Nyquist rate, and the sampled image is later digitized based on the separation of the signal in the frequency spectrum in different bands, considering the parameters like the image resolutions and the number of pixels of an image, etc. The image will have errors, which can be considered almost as a quantization error of digitized signal. Based on the number of binary bits that are generated when digitized in the frequency spectrum, one can choose the required architecture based on the complexity of the task with which one does.

The computations can generate the errors that are negligible or have the least probability of errors only, and the errors and effect of noise need to be reduced. Whenever the noise creeps into the signal, a certain amount of precautions needs to be taken to prevent the casualties due to its presence, like maintaining a certain signal-to-noise ratio and considering the type of noise that one deals with [1–3].

The types of the wavelets are based on the dimensionality for the sake of the computation, and the requirement of the complex task execution and generation of outcomes in an efficient manner are:

- 1-D;
- 2-D;
- 3-D, etc.

The complexity governs the factor of the dimension one must use in order to achieve the outputs. When the image or the signal is represented in the form of the pixels or its components within its frequency spectrum, filter design needs to be as perfect as possible to prevent the aliasing or the generation of the vestigial bands in the signals and, on the whole, affect the shape of the signal and distort it. The design must be as sharp as possible with the least transition band and almost nearer to the ideal frequency response curves. The entire design needs to be operated when only the system is stable and produces no spurious outputs. Then only the effect of the noise can be reduced to a great extent [1–3].

The number of bits or bytes taken for the sake of the computations needs to be considered based on the image resolution and the bit handling capacity of the computational units during a clock cycle. Initially, the data is read serially either bitwise or in a group-wise manner, along the row, and then via a storage element like registers, it is made to send next to the part where there is a column processor, and the computations are done accordingly for the sake of the column, just like the matrix arithmetic is mimicked. The data is added initially, and the carry bit is handled based on the type of the adder that one has chosen in the architecture and then multiplied later with the help of the multiplier units, with the least amount of delay that has been generated by the gates and the carry bit generated needs to be handled carefully at every stage to maintain the accuracy of the entire operation. The noisy data needs to be handled with utmost care in order to prevent the generation of faulty outcomes. The data hence needs to be sent to the processing units (PUs) via multiple numbers of the filter banks such that all the noisy data will be filtered out from the spectrum that has been sampled and digitized along with the selection of the desired frequency components from the main signal in the frequency domain [1–4].

1.2.1   1-DIMENSIONAL ARCHITECTURE

This type of architecture depends on the efficiency of the pipelining at the input data stream in a serial manner. The output depends on the present inputs and the past outputs as well, while the past outputs are fed to the input side with the help of the negative feedback path that has the potential to eliminate the noise or the error that has crept into the signal. The computations are made to be performed via four blocks, namely alpha, beta, gamma, and lambda; these constitute to be the wavelet coefficients. The value of the alpha, beta, gamma, and lambda govern the type of filter that can be designed for one's application requirement. These blocks, along with the help of the adder and the multiplier modules with a certain level of optimal design, are used to implement the architecture. The main advantage of this architecture is that it is easier to implement and design in the reduced area and lower amount of the computations when compared to the higher dimensions. The main disadvantage with this type of architecture is that the entire throughput of the operation that is being performed by it is reduced by it due to no support for the concurrency at the time of the processing of the bits, irrespective of the type of scheme used in order to design the architecture for the sake of the computations. The number of computations that it does is lesser than that of the higher dimensional architectures, but it is quite apt for the simple level of the input data rather than not doing computations over the complex data. It is known to be implemented by an algorithm classically known by the name Mallat's algorithm. In a recursive manner, the implementation can be depicted in the form of a tree. All the lower coefficients are directly calculated with the help of the formula to maintain the ease of the computation (Figure 1.2) [3–5].

The computations are done in obedience to that of the matrix arithmetic implementation, and these computations need to be done via row and the column processors, individually, and the carry bit is handled accordingly wherein the data transfer is done between the processors with the help of the memory bank or registers, to prevent the loss of the data or the corruption of the data based on an operating clock frequency. A memory element is needed to transfer the bits from the row processor to the column processor to compute the computations, where each processor has an adder, and a multiplier unit, and the efficiency of the operation still depends upon the wavelet coefficients, namely alpha, beta, gamma, and lambda. The no of registers that can be used is given by the formula $k = \log_2^N$, where N can be considered as the number of bits that are taken at a single clock frequency for the sake of the computations at the input side. The pipelining of the data plays a major role in determining the delay and speed of the operation of the task [4, 5]. The input data can be either is sent bitwise or it can be sent as a pair or in a group of bits that are to be pipelined in order to increase the throughput of the entire system, reducing the delay and increase the computational speed. The coefficients can also play a major role in the computation of the data. The data is sent to the row processor section wherein the row-wise computations are done and later on to the column processor for column-wise processing or computations via a memory bank, to mimic the matrix arithmetic operations. These tasks must be done in an optimal way to minimize the delay and later to enhance the throughput. Irrespective of the logic family used to design the hardware, all the factors like the operating voltage and the speed of the computational elements must be considered to reduce the delay factor of the entire operation on the whole due to the hardware computational modules. The design must be having lowest area consumption at the small scale and even at the macro scale such that, the entire constraints like the power, speed can also be optimized [1–5, 25–27].

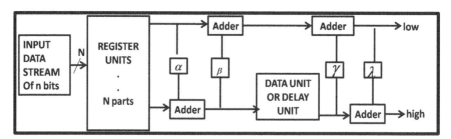

FIGURE 1.2   1-D VLSI architecture [3].

A various implementations can be designed like the Daub-4 and 9/7 filter design by the 1-D architecture for the dwt implementation, which are orthogonal filters. The number of samples taken at the input is greater than the number of samples per wavelet and these wavelets are computed with a certain amount of the latency. The entire set of the wavelets needs to be split into an even and an odd part, where the initial computation consumes more latency for the sake of the feedback path. There are various schemes like lifting, flipping, and convolution, etc., used to implement the design of DWT in VLSI. Each scheme has its own advantages and disadvantages such that one can choose the scheme based on the requirement of the tasks. It is a very basic type of the dimension for the wavelet as the bit handling capacity is lower than compared to that of the higher dimensions (Figure 1.3).

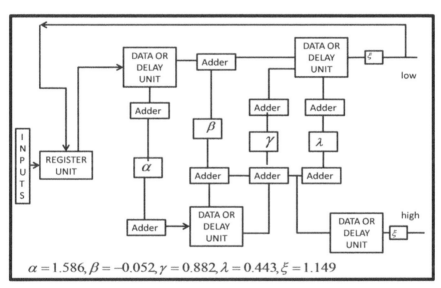

$$\alpha = 1.586, \beta = -0.052, \gamma = 0.882, \lambda = 0.443, \xi = 1.149$$

FIGURE 1.3   9/7 Filter design by 1-D architecture [3].

For example, in the case of the 9/7 filter design, the input samples are separated into an even and an odd parts in the frequency domain by the use of the filter banks and such that by the use of the registers at the input side of the module, at every clock transitioning edges, the odd and the even parts are synchronized to the clock signal and then sent for the computations toward the MAC units (multiplier and adder units) and then later by adjusting the values of the alpha, beta, gamma, lambda, all the coefficients of the wavelets govern the filter design that one requires. These are computed accordingly in order to achieve the desired outcome as these coefficients play a major role over the separated data (Table 1.1) [4, 5, 25–27].

TABLE 1.1   9/7 Filter Computations in 1-D Architecture [4, 5].

SL. No.	Even Part	Odd Part
1.	$P_{0,1} = P_{0,1}$	$P_{0,2} = P_{0,2}$
2.	$\text{Even}_{-1,1} = P_{0,1}$	$\text{Odd}_{-1,1} = \alpha.\, P_{0,1} + P_{0,2}$
3.	$\text{Even}_{-1,1} = \beta\, P_{0,1} + \text{Even}_{-1,1}$	$\text{Odd}_{-1,1} = \text{Odd}_{-1,1}$
4.	$\text{Even}_{-1,1} = z^{-1} \gamma\, \text{Odd}_{-1,1} + \text{Even}_{-1,1}$	$\text{Odd}_{-1,1} = z^{-1} \text{Odd}_{-1,1}$

The implementation is solely dependent on the optimal design of the adder and multiplier units in the row and the column PUs that are being incorporated into the design. The designs must be optimized to reduce the delay at low power consumption and hence achieving a good area and delay product. It must take low power in that small area and must produce very minimal delay with high speed of the computations, maintaining a speed to power trade-off. The speed of the computations depends upon the clock signal frequency for the data to be shifted via the registers between the processors. The row and column computing blocks/ processor must be highly efficient and must take necessary precautions to prevent the throughput to be a bad value and handle the carry bits in an efficient manner. The noise handling capacity can be enhanced further by band-limiting the signal which can be achieved by the use of the various filter banks.

There are various types of adders and multipliers that are available, but the one with the constraints that suit to our requirements needs to be chosen. It can be used to implement the computation on a single bit (bit-serial) or on a group of bits (digit serial) via various methods like Karatsuba multiplication, etc. (Figure 1.4).

1.2.2   2-DIMENSIONAL ARCHITECTURE

It is the repetitive implementation of the 1-D computational procedure and provides a better throughput in terms of the bit handling capacity and the computational process that is being implemented. It can also be implemented by various schemes like lifting, flipping, convolution, etc. It is mostly known to be implemented in the fields like image processing which needs more computational abilities in order to understand or analyze the images at a faster rate when compared to that of the 1-D architecture. It is also backed by the matrix arithmetic. The input end can either take in bitwise or group wise serially or parallel based on the architecture optimal design one has chosen to achieve the outcomes. The memory requirement depends upon the number of bits or the bit handling capacity and is given by the formula $N^2/4$; this memory is placed as an interface between the row and column processor and even in the feedback path. The row and column processor each has its own MAC units to do the computations. The carry bit is handled accordingly based on the type of the adder or multiplier chosen.

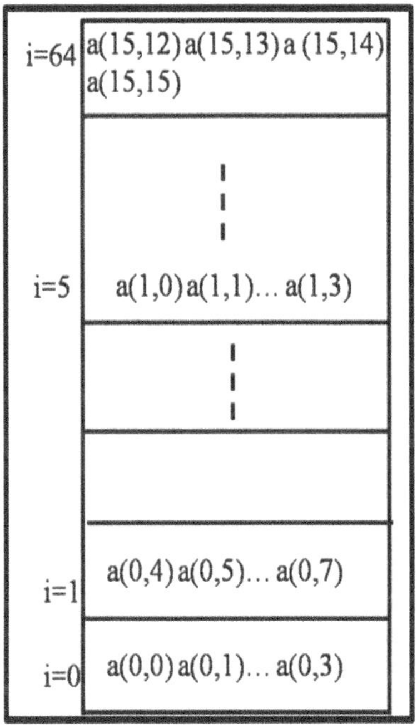

i=64	a(15,12) a(15,13) a (15,14) a(15,15)
	⋮
i=5	a(1,0) a(1,1)... a(1,3)
	⋮
i=1	a(0,4) a(0,5)... a(0,7)
i=0	a(0,0) a(0,1)... a(0,3)

FIGURE 1.4 Data flow order for the computations [3–5].

Its design needs to be optimized to prevent adding any additional delays and power wastages along with the area optimization. The speed of the entire computations depends on the operating voltages or on the clock signal frequency. The architecture must be highly efficient enough to meet our requirements accordingly and have better application in the field of the image processing. The design must be in a position to handle the errors or the noise that is crept into the signal to prevent the generation of the wrong outputs. It can be taken as the repetitive architecture of the 1-D wavelet, and the choice of the coefficient in the 1-D must be taken in a wise manner to do better enhanced computations [3–5]. The value of the coefficients will play a major role in the matrix arithmetic of the data that has been taken such that there will be generation of the output, but the accuracy will be governed by these wavelet coefficients. The taken data is pipelined to have better throughput of the entire operation that is being done, hence a queue needs to be maintained in the architecture to have the pipelining of the data. As the entire efficiency of the operation depends upon the pipelining, there must be a rate at which the data has to be sent in and out of the queue such that, it must be made synchronized o that of the clock signal frequency (Figure 1.5).

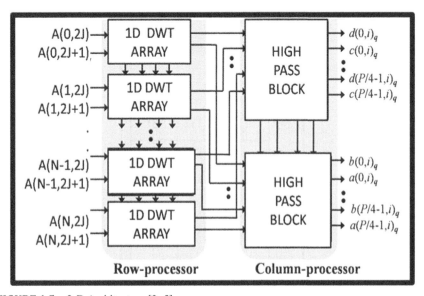

FIGURE 1.5 2-D Architecture [3–5].

The computations can also be implemented by the use of the Radix-2/4/8 methods but any one must be chosen in order to suffice the bit handling capacity and the delay constraint. Unnecessary computations if prevented, it reduces the propagation delay to a larger extent. The input data and the past outputs that are inputted are to be separated accordingly to perform the 1-D computations either in a recursive manner or in a direct scan method. Hence the critical path delays (CPD) irrespective of the scheme that has been used must be handled. There are various modifications being done in the design in order to have eased the computations and at a faster rate.

The computational speed depends on the clock cycle frequency and on the delay in the output generation. It needs to be optimized for the sake of the image when inputted; the processing needs to be done accordingly by selecting the best values of alpha, beta, gamma, and lambda for the individual 1-D implementation. It can be even used for the Daub and other filter designs.

The computation, irrespective of the architectures that are to be implemented, it depends on the signal to noise ratio as well. The SNR ratio prevents the data to be corrupted in the computation and even before it. The effect of the noise can be reduced or minimized only by the help of the band limiting of the signal which is achieved with the help of the bandpass filters. The signal can be sampled and its sampling frequency can be interpolated or decimated accordingly based on the operating conditions. It even depends on the stability of the filters that are being implemented based on the wavelet coefficients accordingly. It must be able to produce outputs in an efficient manner with minimal errors.

1.2.3   3-DIMENSIONAL ARCHITECTURE

The computation of the DWT can be done in the 3-dimensional methods such that all image processing can be done with an ease which are complex to analyze. The data is compressed in the temporal and spatial domains, and the data is processed by the use of the structural PU and a row and a column processor. The data processing will be achieved by the schemes like the lifting and convolution methods, and the design needs to be optimized to a large extent. It basically consumes more area and hence the operating conditions needs to be enhanced to such an extent that the throughput should become an optimized one. It takes more computing elements like the adders and the multipliers, and its functionality can be tested on a field-programmable gate array (FPGA) as it is quite apt for the DSP applications. It can be considered as the repetitive version of the 2-dimensional or the 1-Dimensional architecture. Due to this repetitive architecture, the computations can be done for the large and complex data. The adder and multiplier units of the MAC play a major role in the computations of the 1-Dimensional data or the 2-dimensional data. The delay got at the end of the computations must be reduced to a larger extent such that speed of the entire operation can be enhanced to a greater extent.

The design can be used for the various image processing applications, and the hardware design needs to be optimized considering the operating conditions and the memory and clock frequency requirements. The input will be divided into the sub-bands namely LL, LH, HL, HH, such that all the data which must be sent for the processing must be pipelined or grouped accordingly based on the requirement one has for the complexity of the data. The data will be divided into bits and then sent to the temporal and spatial processors. The CPD of the methods needs to be considered in order to optimize the design. The output of the spatial processors is given to the input of the temporal processors. Filters like the CDF 9/7 can be designed such that they can be used for the discrete wavelet applications. A data flow graph can be used for the sake of better understanding of the operation. For every 2 clock cycles, the computations are done (Figure 1.6).

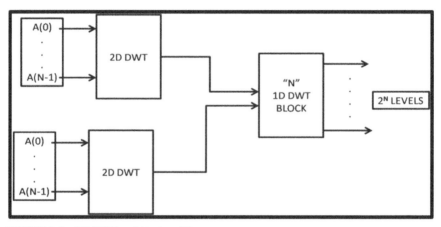

FIGURE 1.6   3D DWT architecture [8].

1.3   SCHEMAS FOR THE DWT IMPLEMENTATION

There are various schemes which are used to implement the DWT at the time of the designing. The most basic techniques are lifting, flipping, and convolution that are used to implement the design. Each method has its own advantages and disadvantages that govern the factors like throughput, delay, computation speed, efficiency of the filters being used. Each has its own type of the architecture and varies with each other in the implementation. All the schemas share a similar amount of the architectures and have various amount of the PUs that can generate the outputs at a certain level of the throughput. These methods depend upon the value of the coefficients, namely alpha, beta, gamma, lambda chosen for the computations to be done in an optimal and efficient way. The architectures can be realized by any logic family one wishes to implement based on the operating voltage conditions and the operating speed one desires such that the outcomes can be generated at a better rate. The basic logic families like the bipolar junction transistor or the field effect transistors can be used in the design based on the person's requirements of the area, power, and speed. The mostly preferred transistors at the nanoscale are the metal oxide semiconductor field-effect transistors (FETs) named as the MOSFET's. The CMOS logic of the MOSFET's can be used to realize the architecture physically as they have large packing densities and lesser operating voltage conditions with faster speed of the operation than the bipolar junction transistor (Figures 1.7 and 1.8).

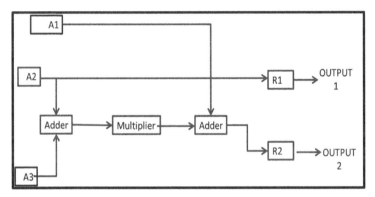

FIGURE 1.7   Basic lifting cell architecture.

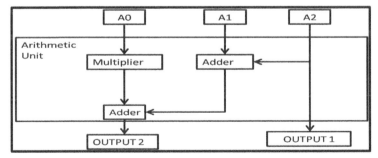

FIGURE 1.8   Basic flipping cell architecture.

1.3.1   *LIFTING SCHEME*

It has basically three steps namely:
- Split;
- Predict; and
- Update.

In this scheme, the entire samples are divided to form a set of an even and odd parts such that the values are predicted and updated at the end of every computation. The computations obey the matrix arithmetic. The upper triangle of the matrix has the elements to predict, and the lower triangle of the matrix has the values that are to be updated. These iterations or the multiple scans that are done helps to update the matrix and predict the next values for the sake of the further computations. The values when added and multiplied accordingly, a polynomial equation is formed that are used to analyze the entire operation of the system. The data must be sent to the column processing part after the computation in the row module has been completed. The value of the coefficients plays a major role in the computations (Figure 1.9) [6, 7, 25].

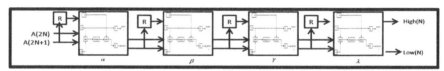

FIGURE 1.9   A general lifting architecture for 1D DWT.

It is used in the design of the 9/7 filter bi-orthogonal wavelet filter wherein which is presently used in the present-day implementations for the sake of the noise and error removal and ease of computations. There is a coefficient k or f1 used for the sake of the scaling of the signal which is given to it to be filtered (Figure 1.10) [7].

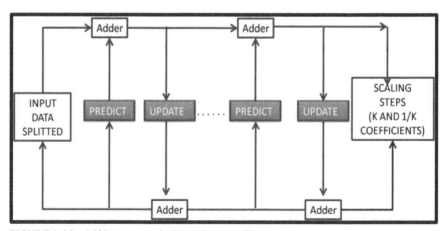

FIGURE 1.10   Lifting process in 9/7 orthogonal filter-1.

It is achieved by the help of the filter banks that are being used. These filter banks help in the splitting of the data into an even and the odd parts in the frequency domain, due to the division of the spectrum into the sub-bands, the data is bandlimited and can have a good signal to noise ratio and then reduce the effects of the noise. Later the values are predicted and later updated in the last 2 stages, respectively, wherein with the help of the alpha, beta, gamma, lambda values, the coefficients of the wavelets can be determined.

Let,

$$P_0(j) = A(2j+1) \tag{1.3}$$

$$Q_0(j) = A(2j) \tag{1.4}$$

Then,

$$P_1(j) = P_0(j) + \alpha * \left[Q_0(j) + Q_0(j+1)\right] \tag{1.5}$$

$$Q_1(j) = Q_0(j) + \beta * \left[P_0(j-1) + P_0(j)\right] \tag{1.6}$$

$$P_2(j) = P_1(j) + \gamma * \left[Q_1(j) + Q_1(j+1)\right] \tag{1.7}$$

$$Q_2(j) = Q_1(j) + \lambda * \left[P_2(j-1) + P_2(j)\right] \tag{1.8}$$

Therefore,

$$High(j) = \left(\frac{1}{f1}\right) * P_2(j) \tag{1.9}$$

$$Low(j) = (f1) * Q_2(j) \tag{1.10}$$

It can be even implemented with or without pipelining feature which may reduce the throughput based on the size of the data with which one can handle. The pipelining may result in increasing the number of registers that but the area can be optimized and efficiency can be decreased, hence in order to improve the overall efficiency of the operation or the speed of the operation, one must use the feature of the pipelining. The design of the adder and the multiplier that are being used needs to be as optimized as possible to reduce the effects of the computational delay. The delay must be as low as possible in order to do the computational tasks. The main disadvantage is the problem of the CPD needs to be handled due to the presence of the adders and multiplier even though if they have an optimal design. As the incorporation is done in a very small scale or at the nanoscale, the effect of the parasitic must also be handled as they tend to creep into the architecture at a certain frequency or at smaller areas. The parasitic mostly tend to cause the delay and hence units like the buffers needs to be used in order to increase the strength of the signal [6–8].

1.3.2   FLIPPING SCHEME

Flipping is also a method is used to implement the DWT design. It is much more efficient than the lifting technique as the problem of the CPD has been reduced. The use of the Radix-4 multiplier reduces the CPD to a great extent and it increases the throughput as well positively. The computations are done in a quick manner and the entire delay depends upon the delay of the multiplier and the adder only. An encoder also needs to be used to encode the values accordingly, and the values of the alpha, beta, gamma, and lambda are selected accordingly for the implementation (Figure 1.11) [6–8].

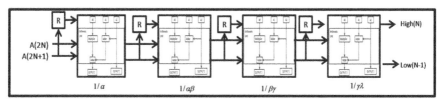

FIGURE 1.11   Flipping architecture for 1-D DWT.

It has its own disadvantages in the long run due to the reduced use of the registers, and that may cause the glitches or the errors to creep into the signal that may affect the signal even though its throughput may be higher than the lifting scheme. It can even be considered as multiple implementations of the lifting schemas at once that has optimal throughput. They can be tested in the FPGA irrespective of the dimension one requires; it can be used very much aptly for the sake of the digital data processing and computing of the data with the same amount of the lesser throughput one desires. The area-delay product (ADP) must be giving value indicating that the scheme can be found ideal to do any sort of the computations at a faster rate, but the disadvantage of this method is that it can be easily susceptible to the errors.

$$y(N) = a(2N) \tag{1.11}$$

$$y_1(N) = a_1 A(2N+1) + \left[A(2N) + A(2N+2)\right] \tag{1.12}$$

$$y_2(N) = a_2 A(2N) + [y_1(N-1) + y_1(N)] \tag{1.13}$$

$$high(N) = K_1\left[a_3 y_1(N) + \left[y_2(N) + y_2(N+1)\right]\right] \tag{1.14}$$

$$low(N) = K_2\left[a_4 y_2(N) + \left[\left[high(N-1)/K_1\right] + \left[high(N)/K_2\right]\right]\right] \tag{1.15}$$

1.3.3 CONVOLUTION SCHEME

This type of implementation may seem to have a greater number of the computations, but there is no requirement of the temporary registers to store the data, and hence the design becomes much more efficient in terms of the area and power and speed considerations. The design can be even be optimized by not using a multiplier and rather a shift register can be used and addition of the bits can be done accordingly by shifting of the bits. This can be achieved by the use of the shift registers that shift the bits in order to do the computations that are very optimal enough to achieve the desired outcome. The method can have the large usage of the filter banks such that, all the data can be given for the processing and generate the outcomes accordingly.

The architecture of the convolution has a row and column processor that requires more arithmetic resources than the lifting scheme in order to generate the outputs. The architecture can be used to design various filters like the 9/7 or 5/3 orthogonal filters, and it can also be used with the JPEG 2000 compression method accordingly that can do the computations in a much efficient way which is optimized and gives an area-delay-product

(ADP) as an optimized one. It requires a certain number of memory banks that can be used either to store the shifted bits or transfer the bits from the column to row processor accordingly, hence the design may give a lower throughput to visualize but the entire throughput will be depending upon the clock cycle frequency and voltage at which it can be operated. The design can be verified for the sake of the testing in a FPGA, as it is quite good enough to be used for the DSP applications (Figure 1.12).

A basic two-dimensional convolution scheme of DWTs is given by:

$$\text{OUTPUT}_{j+1}\left(N_1, N_2\right) = \sum_{j=0}^{L-1}\sum_{j=0}^{L-1} A\left(N_1, N_2\right).F_j\left(2N_1 - K_1, 2N_2 - K_2\right) \qquad (1.16)$$

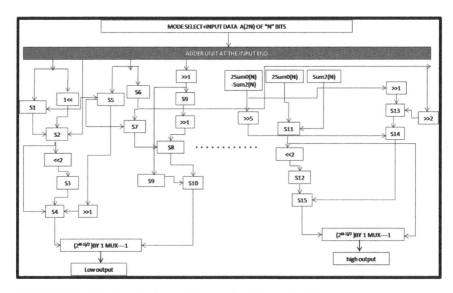

FIGURE 1.12   1-D Convolution architecture for 9/7 or a 5/3 filter.

The convolution architecture generally has a larger number of the computations when compared to that of the other schemas as there are many additions and multiplications being performed at each and every stage. The area can be optimized with the help of the removal of the multiplier units and replacing it with the set of the shift registers and adder units and hence achieving the same output. The multiplication can be mimicked by shifting the bits by a certain amount of desired number and then by the use of the adder units, the computations are done accordingly.

Similarly, in the case of the toe dimensional architecture, it can be considered as a repetitive architecture of the one-dimensional architecture wherein, the convolutions can be done with the help of the successive corresponding adder and multiplier units. This type of design will have proper utilization of every resource block that has been used in the architecture such that, the number of computations will increase but the throughput can be considered moderate and quite lesser when compared to that of the flipping scheme but almost equal to that of the lifting scheme. Irrespective of the operating voltage conditions, the speed or the delay of the entire task will depend upon the number of computations done at every stage and on the clock frequency to shift the data. The row and column processor will have a greater number of computations, but then it can be generating results at a slower rate when compared to that of the flipping scheme of DWT as there are large numbers of computations.

1.4 SUMMARY OF THE ARCHITECTURES

Of all the schemes, the lifting is preferred for its moderate throughput and its optimal design whereas, 2-dimensional is preferred for the sake of the image processing applications. There are various ranges of the values of the number of the adder and the Multiplier units used in the design, and they vary based on the requirement of the designer's design optimization capabilities (Table 1.2).

TABLE 1.2 Summary of All the Architectures of DWT

SL. No.	Type of DWT Architecture	MAC Units	Memory Banks	Delay (ns)	Schemes Preferably Applicable
1.	1-Dimensional	2–20	5–10	5–20	
2.	2-Dimensional	10–25	8–25	10–25	Lifting, flipping, convolution
3.	3-Dimensional	20–40	10–45	10–45	

1.5 CONCLUSION

The DWT can be implemented in the VLSI using any Hardware definition language in the front-end design or via transistor-to-transistor connection in the back-end full custom design. The design has to be optimized every time in accordance to the advancements in the methods and the technologies.

The computations are governed by considering the area-delay-product (ADP) for the IC design. Any signal when divided into sub-bands within its spectrum and then sampled and digitized, it helps to have an ease of computations, saving power and having the same accuracy as compared to that of the continuous frequency domain, for the results in order to have the analysis of the image. Of all the schemes, the lifting scheme is found suitable for the operations even though the throughput of the process is found to be lesser than that of the other schemas and the architecture for the 2-D wavelets is much more useful for the sake of the image processing. The preferable scheme will be that lifting scheme in the 2-dimensional architecture design that finds the majority of the application in the field of the image processing.

It has many applications in the field of the medical, nuclear, space, and defense and research sectors as well, where it majorly depends upon the images that are to be processed to derive the meaningful data from the input data that has been given to it. Due to the advent of the artificial intelligence and deep learning, these types of the computations can be made much easier and in a much more effective manner, which can play a major role in generating the outcomes of the required task. A Neural Network when developed in order to process the image, it can give its own decision accordingly at the end of the completion of the outcome and hence one can find it much more advantageous in the mere future to incorporate this type of hardware in the other designs. Hence, DWT plays a major role in the understanding of the image with an ease and with enhancement of the proposed design in terms of ADP product, every time the efficiency of the entire task can be done. The overall efficiency of the operation can also depend upon the parasitic of the transistors used in order to design the gates in the multiplier and adder units, hence precautions are to be taken.

KEYWORDS

- **convolution**
- **critical path delay**
- **discrete wavelet transform**
- **flipping**
- **lifting**
- **MAC unit**

REFERENCES

1. Tabassum, F., Islam, M. I., & Amin, M. R., (2021). Comparison of filter banks of DWT in recovery of image using one dimensional signal vector. *Journal of King Saud University - Computer and Information Sciences.* https://doi.org/10.1016/j.jksuci.2019.03.005.

2. Hasan, D., & Gholamreza, A., (2011). Discrete wavelet transform-based satellite image resolution enhancement. *IEEE Transactions on Geoscience and Remote Sensing, 49*(6), 1997–2004.

3. Hongyu, L., Mrinal, K. M., & Bruce, F. C., (2004). Efficient architectures for 1-D and 2-D lifting-based wavelet transforms. *IEEE Transactions on Signal Processing, 52*(5), 1315–1326.

4. Wei, Z., Zhe, J., Zhiyu, G., & Yanyan, L., (2012). An efficient VLSI architecture for lifting-based discrete wavelet transform. *IEEE Transactions on Circuits and Systems-II: Express Briefs, 59*(3), 1080–1089.

5. Chao-Tsung, H., Po-Chih, T., & Liang-Gee, C., (2005). Analysis and VLSI architecture for 1-D and 2-D discrete wavelet transform. *IEEE Transactions on Signal Processing, 53*(4).

6. Chengjun, Z., Chunyan, W., & Omair, A. M., (2010). A pipeline VLSI architecture for high-speed computation of the 1-D discrete wavelet transform. *IEEE Transactions on Circuits and Systems—I: Regular Papers, 57*(10), 2729–2740.

7. Tinku, A., & Chaitali, C., (2006). A survey on lifting-based discrete wavelet transform architectures. *Journal of VLSI Signal Processing, 42*, 321–339. doi: 10.1007/s11266-006-4191-3.

8. Srinivasrao, B. K. N., & Indrajit, C., (2016). *High Performance VLSI Architecture for 3-D Discrete Wavelet Transform.* 978-1-4673-9498-7/16/$31.00 C20016 IEEE.

9. Chakraborty, D., & Banerjee, A., (2017). *A Multiplier Less VLSI Architecture of Modified Lifting Based 1D/2D DWT Using Speculative Adder.* 978-1-5090-3800-8/17/$31.00 ©2017 IEEE.

10. Ganapathi, H., Kotha, S. R., & Telugu Kuppu, S. R., (2018). *A New Approach for 1-D and 2-D DWT Architectures Using LUT Based Lifting and Flipping Cell.* https://doi.org/10.1016/j.aeue.2018.10.002 1434-8411/_2018 Elsevier GmbH. All rights reserved.

11. Abdul, K., & Mohammed, B., (2018). Design and development of multimodal medical image fusion using discrete wavelet transform. In: *2018 Second International Conference on Inventive Communication and Computational Technologies (ICICCT)* (pp. 1629–1633).

12. Ahammed, K., Ershadullah, M., Rakebul, I. H. M., & Islam, S., (2015). Design and implementation of digital filter bank to reduce noise and reconstruct the input signals. *Signal Image Process (SIPIJ), 6*(2), 15–18.

13. Al Jumah, A., (2013). Denoising of an image using discrete stationary wavelet transform and various thresholding techniques. *J. Signal Inf. Process, 2013*(4), 33–41.

14. Arai, K., & Rahmad, C., (2012). Wavelet based image retrieval method. *Int. J. Adv. Comput. Sci. Appl., 3*(4), 6–11.

15. Wenkai, T., Huapeng, W., & Ahmadi, M., (2005). VLSI implementation of bit-parallel word-serial multiplier in GF(2/sup 233/). *The 3rd International IEEE-NEWCAS Conference, 2005* (pp. 399–402). Quebec City, Que.doi: 10.1109/NEWCAS.2005.1496706.

16. Chen, Z., Hao, H., Li, L., & Dong, J., (2006). Wavelet transform for rabbit EEG with vagus nerve electric stimulation. In: *2006 International Conference of the IEEE Engineering in Medicine and Biology Society* (pp. 1715–1718). New York, NY. doi: 10.1109/IEMBS.2006.260698.

17. Suhail, M. A., & Obaidat, M. S., (2001). On the digital watermarking in JPEG 2000. In: *8th IEEE International Conference on Electronics, Circuits and Systems (Cat. No.01EX483)* (Vol. 2, pp. 871–874). Malta. doi: 10.1109/ICECS.2001.957612.

18. Christopoulos, C., Skodras, A., & Ebrahimi, T., (2000). The JPEG2000 still image coding system: An overview. In: *IEEE Transactions on Consumer Electronics* (Vol. 46, No. 4, pp. 1103–1127). doi: 10.1109/30.920468.

19. Chung-Neng, W., Tihao, C., Chi-Min, L., & Hung-Ju, L., (2001). Improved MPEG-4 visual texture coding using double transform coding. *The 2001 IEEE International Symposium on Circuits and Systems (Cat. No.01CH37196)* (Vol. 5, pp. 227–230). Sydney, NSW. doi: 10.1109/ISCAS.2001.922026.

20. Hassen, W., & Amiri, H., (2013). The 5/3 and 9/7 wavelet filters study in a sub-bands image coding. In: *2013 7th IEEE International Conference on E-Learning in Industrial Electronics (ICELIE)* (pp. 150–154). Vienna. doi: 10.1109/ICELIE.2013.6701290.

21. Cohen, A., Daubechies, I., & Feauveau, J. C., (1992). Biorthogonal bases of compactly supported wavelets. *Comm. Pure Appl. Math., 45*, 485–560. doi: 10.1002/cpa.3160450502.

22. Wim, S., (1996). *The Lifting Scheme: A Custom-Design Construction of Biorthogonal Wavelets, Applied and Computational Harmonic Analysis, 3*(2), 186–200. ISSN 1063-5203. https://doi.org/10.1006/acha.1996.0015.

23. Calderbank, A. R., Ingrid, D., Wim, S., & Boon-Lock, Y., (1998). Wavelet transforms that map integers to integers. *Applied and Computational Harmonic Analysis, 5*(3), 332–369. ISSN 1063-5203. https://doi.org/10.1006/acha.1997.0238.

24. Liu, C. C., (2000). Design and implementation of a progressive image coding chip based on the lifted wavelet transform. In: *Proc. 11th VLSI Design/CAD Symposium,* (pp. 49–52).

25. Chung-Jr, L., Kuan-Fu, C., Hong-Hui, C., & Liang-Gee, C., (2001). Lifting based discrete wavelet transform architecture for JPEG2000. *The 2001 IEEE International Symposium on Circuits and Systems (Cat. No. 01CH37196)* (Vol. 2, pp. 445–448). Sydney, NSW. doi: 10.1109/ISCAS.2001.921103.

26. Wei-Hsin, C., Yew-San, L., Wen-Shiaw, P., & Chen-Yi, L., (2001). A line-based, memory efficient and programmable architecture for 2D DWT using lifting scheme. *The 2001 IEEE International Symposium on Circuits and Systems (Cat. No. 01CH37196)* (Vol. 4, pp. 330–333). Sydney, NSW. doi: 10.1109/ISCAS.2001.922239.

27. Chao-Tsung, H., Po-Chih, T., & Liang-Gee, C., (2004). Flipping structure: An efficient VLSI architecture for lifting-based discrete wavelet transform. In: *IEEE Transactions on Signal Processing* (Vol. 52, No. 4, pp. 1080–1089). doi: 10.1109/TSP.2004.823509.

28. Andra, K., Chakrabarti, C., & Acharya, T., (2002). A VLSI architecture for lifting-based forward and inverse wavelet transform. In: *IEEE Transactions on Signal Processing* (Vol. 50, No. 4, pp. 966–977). doi: 10.1109/78.992147.

29. Dillen, G., Georis, B., Legat, J. D., & Cantineau, O., (2003). Combined line-based architecture for the 5-3 and 9-7 wavelet transform of JPEG2000. In: *IEEE Transactions on Circuits and Systems for Video Technology* (Vol. 13, No. 9, pp. 944–950). doi: 10.1109/TCSVT.2003.816518.

30. Bing-Fei, W., & Chung-Fu, L., (2003). A rescheduling and fast pipeline VLSI architecture for lifting-based discrete wavelet transform. *Proceedings of the 2003 International Symposium on Circuits and Systems* (pp. 732–735). Bangkok. doi: 10.1109/ISCAS.2003.1206078.

31. Silva, S. V., & Bampi, S., (2005). Area and throughput trade-offs in the design of pipelined discrete wavelet transform architectures. *Design, Automation and Test in Europe, 3*, 32–37. Munich, Germany. doi: 10.1109/DATE.2005.66.

32. Martina, M., & Masera, G., (2006). Low-complexity, efficient 9/7 wavelet filters VLSI implementation. In: *IEEE Transactions on Circuits and Systems II: Express Briefs* (Vol. 53, No. 11, pp. 1289–1293). doi: 10.1109/TCSII.2006.883092.

33. Meher, P. K., Mohanty, B. K., & Swamy, M. M. S., (2015). Low-area and low-power reconfigurable architecture for convolution-based 1-D DWT using 9/7 and 5/3 filters. In: *2015 28th International Conference on VLSI Design* (pp. 327–332). Bangalore. doi: 10.1109/VLSID.2015.61.

34. Mohanty, B. K., & Meher, P. K., (2011). memory-efficient modular VLSI architecture for high throughput and low-latency implementation of multilevel lifting 2-D DWT. In: *IEEE Transactions on Signal Processing* (Vol. 59, No. 5, pp. 2072–2084). doi: 10.1109/ TSP.2011.2109953.

35. Mohanty, B. K., Mahajan, A., & Meher, P. K., (2012). Area- and power-efficient architecture for high-throughput implementation of lifting 2-D DWT. In: *IEEE Transactions on Circuits and Systems II: Express Briefs* (Vol. 59, No. 7, pp. 434–438). doi: 10.1109/TCSII.2012.2200169.

36. Mohanty, B. K., & Choubey, A., (2017). Efficient design for radix-8 booth multiplier and its application in lifting 2-D DWT. *Circuits Syst Signal Process, 36*, 1129–1149. https://doi.org/10.1007/s00034-016-0349-9.

37. Mohanty, B. K., Meher, P. K., & Srikanthan, T., (2015). Critical-path optimization for efficient hardware realization of lifting and flipping DWTs. In: *2015 IEEE International Symposium on Circuits and Systems (ISCAS)* (pp. 1186–1189). Lisbon. doi: 10.1109/ ISCAS.2015.7168851.

38. Meher, P. K., Mohanty, B. K., & Patra, J. C., (2008). Hardware-efficient systolic-like modular design for two-dimensional discrete wavelet transform. In: *IEEE Transactions on Circuits and Systems II: Express Briefs* (Vol. 55, No. 2, pp. 151–155). doi: 10.1109/ TCSII.2007.911801.

39. Choubey, A., & Mohanty, B. K., (2018). Novel data-access scheme and efficient parallel architecture for multilevel lifting 2-D DWT. *Circuits Syst Signal Process, 37*, 4482–4503. https://doi.org/10.1007/s00034-018-0775-y.

40. Beth, T., & Gollmann, D., (1989). Algorithm engineering for public-key algorithms. *IEEE Journal on Selected Areas in Communications, 7*(4), 458–466.

41. Guo, J. H., Wang, C. L., (1998). Digit-serial systolic multiplier for finite fields GF(2m). *IEE Proc. Comput. Digital Tech., 145*(2), 143–148. ISSN: 1359-7027. doi: https://doi. org/10.1049/ip-cdt:19981906.

42. Hankerson, D., Menezes, A. J., & Vanstone, S., (2006). *Guide to Elliptic Curve Cryptography*. Springer Science & Business Media.

43. Hariri, A., & Reyhani-Masoleh, A., (2011). Digit-level semi-systolic and systolic structures for the shifted polynomial basis multiplication over binary extension fields. In: *IEEE Transactions on Very Large-Scale Integration (VLSI) Systems* (Vol. 19, No. 11, pp. 2125–2129). doi: 10.1109/TVLSI.2010.2066994.

44. Rodriguez-Henriguez, F., & Koc, C. K., (2003). Parallel multipliers based on special irreducible pentanomials. In: *IEEE Transactions on Computers* (Vol. 52, No. 12, pp. 1535–1542). doi: 10.1109/TC.2003.1252850.

45. Hsu, I. S., Truong, T. K., Deutsch, L. J., & Reed, I. S., (1988). A comparison of VLSI architecture of finite field multipliers using dual, normal, or standard bases. In: *IEEE Transactions on Computers* (Vol. 37, No. 6, pp. 735–739). doi: 10.1109/12.2212.

46. Imana, J. L., Sanchez, J. M., & Tirado, F., (2006). Bit-parallel finite field multipliers for irreducible trinomials. In: *IEEE Transactions on Computers* (Vol. 55, No. 5, pp. 520–533). doi: 10.1109/TC.2006.69.

47. Jain, S. K., Song, L., & Parhi, K. K., (1998). Efficient semi-systolic architectures for finite-field arithmetic. In: *IEEE Transactions on Very Large-Scale Integration (VLSI) Systems* (Vol. 6, No. 1, pp. 101–113). doi: 10.1109/92.661252.

48. Chang, H. K., Chun, P. H., & Soonhak, K., (2005). A digit-serial multiplier for finite field GF(2/sup m/). In: *IEEE Transactions on Very Large-Scale Integration (VLSI) Systems* (Vol. 13, No. 4, pp. 476–483). doi: 10.1109/TVLSI.2004.842923.

49. Lee, C. Y., (2003). Low complexity bit-parallel systolic multiplier over GF (2m) using irreducible trinomials. *IEE Proceedings-Computers and Digital Techniques, 150*(1), 39–42. doi: https://doi.org/10.1049/ip-cdt:20030061.

50. Lee, C. Y., Horng, J. S., Jou, I. C., Lu, E. H., (2005). Low complexity bit-parallel systolic Montgomery multipliers for special class of *GF(2m)*. *IEEE Trans. Comput., 54*(9), 1061–1070.

51. Lidl, R., & Niederreiter, H., (1994). *Introduction to Finite Fields and Their Applications*. Cambridge university press.

52. Dahab, R., & López, J., (2000). *An Overview of Elliptic Curve Cryptography*. Institute of Computing State University of Campinas Brazil, doi=10.1.1.37.2771.

53. Reyhani-Masoleh, A., & Hasan, M. A., (2004). Low complexity bit parallel architectures for polynomial basis multiplication over GF(2m). In: *IEEE Transactions on Computers* (Vol. 53, No. 8, pp. 945–959). doi: 10.1109/TC.2004.47.

54. Meher, P. K., (2008). Systolic and super-systolic multipliers for finite field $GF(2^{m})$ based on irreducible trinomials. In: *IEEE Transactions on Circuits and Systems I: Regular Papers* (Vol. 55, No. 4, pp. 1031–1040). doi: 10.1109/TCSI.2008.916622.

55. Meher, B. K., & Meher, P. K., (2013). An efficient lookup table-based approach for multiplication over GF(2^m) generated by trinomials. *Circuits Syst Signal, 32*, 2623–2638. https://doi.org/10.1007/s00034-013-9553-z.

56. Scott, P., Tavares, S., & Peppard, L., (1986). A fast VLSI multiplier for GF(2m). In: *IEEE Journal on Selected Areas in Communications* (Vol. 4, No. 1, pp. 62–66). doi: 10.1109/JSAC.1986.1146305.

57. Song, L., & Parhi, K. K., (1998). Low-energy digit-serial/parallel finite field multipliers. *The Journal of VLSI Signal Processing-Systems for Signal, Image, and Video Technology, 19*, 149–166. https://doi.org/10.1023/A:1008013818413.

CHAPTER 2

Execution of Lifting-Scheme Discrete Wavelet Transform by Canonical Signed Digit Multiplier

GUNDUGONTI KISHORE KUMAR[1] and NARAYANAM BALAJI[2]

[1]Department of Electronics and Communication Engineering,
V. R. Siddhartha Engineering College, Vijayawada,
Andhra Pradesh – 520007, India, E-mail: kishore.chiya@gmail.com

[2]Department of Electronics and Communication Engineering,
JNTUK University College of Engineering, Kakinada,
Andhra Pradesh – 533003, India

ABSTRACT

The behavior of the multiplier greatly affects the usefulness of the frame level, especially in image preparation and signal handling applications. In this observation, multiplexers based on the sum-of-power-of-two (SOPOT)-based are investigated, and some exaggerated converters and multiplexers are seen from this examination. Useless "shifters" and "multiplexers" present in "SOPOT" engineering are eliminated. A new architecture of multiplier is proposed that uses the canonical-signed digit (CSD) representation and the control register size is reduced. Lean engineering is better than the current design because of field reduction and use of force. The height scheme discrete wavelet transform (DWT) is executed using an adjusted multiplier. Combination results suggest that the transformed CSD area delay product (ADP) deferred element and vitality are not individually planned above.

2.1 INTRODUCTION

In the mathematical assessment and practical investigation, a "discrete wavelet transforms (DWT)" is several wavelets modify for which the wavelets are unconnectedly examined. Similarly, by means of supplementary wavelet transforms, a key preferred position it has over Fourier changes been the knowing goal: it catches together repetition and area data.

Hungarian mathematician Alfred Haar designed the primary DWT. The Haar wavelet transition could be judged to be equal to input values for statistics spoken to by a ramshackle of numbers, putting away the dissimilarity and passing the sum. This loop is reworked, merging all of the components to show the following scale, creating $2n-1$ contradictions and a final completion. In 1988, Ingrid Daubechies, a Belgian mathematician, spoke on the most used spectrum of discrete wave shifts. This term is based on repeat relationships in order to build a variety of distinct instances. Each goal is twice the previous size, which is dynamically superior to the expected wave. Daubechies describes a category of waves in his original paper, the first of which is a wave of loss. Then excitement for the region grew considerably and a lot of waves were produced that are unique to Daubechies. There has been a broad utilization of DWT in numerous uses of picture preparing and signal handling [1, 2]. The current plans for the execution of the DWT are generally sorted into two kinds: 9/7 channel and the lifting plan. For the subsequent class plans, have points of interest when contrasted with the main classification structures in zone necessity and processing intricacy.

The "digital signal processing (DSP)" applications, for example, "discrete wavelet" change "finite impulse response (FIR) filter" and "infinite impulse response (IIR) filter" includes increase and expansion as central activities. In the constant condition having equipment limitations, deferral, region, and force measurements assumes a key function in dissecting the framework execution. The multiplier assumes key function as equipment perspective in discrete wavelet change, as result multiplier devours bigger region, force, and deferral.

DWT is an intelligent time and frequency analysis method with multi-resolution. The DWT provides higher cryptographic potency, magnitude relationship of compression and image restore value than the usual separate normal feature. As a result, compression formats such as "MPEG-4" and "JPEG 2000" supports the DWT. In signal cryptography [8], the separate moving ridge remodel (DWT) has gained broad unfold acceptance of information compression, information perception, information activity, geophysics, meteorology, audio signal system, motion pursuit, and machine learning (ML). Area units are especially suitable when measurability and tolerable degradation are required for applications. In contrast to a separate circular function remodel, DWT has higher compression ratios, no obstruction artifacts, and clever frequency and time domain localization inherent in scaling and greater flexibility. Several well-known DWT architectures based on a convolution are poorly designed for large scale (VLSI) integrations. The lifting-based DWT was then extracted by Daubechies and Sweden to cut advanced calculations. DWT based on the elevator has many advantages compared to DWT based on the turbocharged method, along with fast implementation, maximum number coefficients, completely reduced quality of hardware and isobilateral forward and backward remodeling. As a consequence of the lifting method, some multipliers will save on scaling operations, integrally a square measure that has a higher CPD than the flip structure. It provides a considerably good structure for area delays rather than a flipping structure. The projected reversal structure is a critical DWT design to minimize the vital path by eliminating multipliers from the input node to the computing node, though not the overhead hardware. By integrating the associate prediction in the nursing update process, the primitive knowledge path of lifting is updated to generate an economic pipeline design that has the fundamental path of one number. The scaling constants of the theme square test excellent opposite of all different, while the scaling constants of the flipping theme do not work so well. Despite that, lifting is thus not normal, just like the turn-around issue, because of the longer critical path, to obtain area-delay economical hardware efficient structures for the "2D-DWT."

Use high efficiency memory architecture in parallel scanning methods for lifting 2D DWT architecture [13]. The first step is to save input and 9/7 filter coefficients with the 4N temporal memory. 4N +8P on-chip memory words are needed without using data access registries and the lifting method "2D-DWT" architecture [12]. Another lift data flow [11] that uses parallel pipeline operations with no effect on critical paths to decrease the symmetric architecture data path. A DWT [10] to transform "serial-parallel" data based on lifting and then sent to the column filters to generate four subgroups, which are then given to transpose the unit to fulfill the order of data flow needed by a row filter.

In several mathematical operations, multiplication is one of the main fundamental functions. One of the most common operations of arithmetic, Fourier's fast and low-performance applications have a dominating role to play in various applications, such as FIR-filters, DWT, Fourier Transform, etc. In the DSP system, the time delays are normal, and the silicone area in the DSP system is very high. The selection of a high-speed multiplier is acceptable because execution time is dominant in many DSP algorithms.

Multiplier is the arithmetic unit performing multiplication between given numbers and one of the factors deciding the consumption of critical path delay (CPD) and chip area. Therefore, the need of designing a multiplier with less area, low power and fast processing makes a demand for effective implementation of design. Several multiplier designs are developed and among them Booth multiplier is known best for its sign implementation and design of hardware. For reducing a number of partial product arrays (PPAs) in Booth multipliers [9], different encoding schemes for Booth algorithm are presented in modified methods.

A regular and modular bit-parallel (word-level) computing structure for DWT easily obtained using multiplier. However, multiplier when implemented in dedicated hardware involves significantly more combinational logic than the adder. A parallel structure of DWT involves a large number of multipliers. Consequently, multiplier consumes a significant part of the chip "area and power." Numerous multipliers can be stored in the parallel design and the throughput rate depends on the availability of combinational resource in the chip. Silicon-area, speed, power consumption is considered the general performance evaluation parameters while designing a VLSI architecture. Previously, energy consumption was secondary to designers than space and speed. However, in recent years, the growth of portable and wireless multimedia devices has made them comparable or even heavier with location and speed. The most significant problem in the architecture of these devices is average energy consumption [14]. As high-performance wireless and portable devices became ubiquitous, battery technology has reached its limits. Energy-efficient designs are more attractive. There are various factors that contribute to the power budget of a VLSI circuit. Combinational logic complexity is one of them. Therefore, low-complexity design most often leads realization of a low-power circuit. Several schemes and methodologies have been proposed and found in the literature for low-complexity design.

Therefore, a whole range of multipliers with very extraordinary zone speed requirements is planned with completely parallel. Multipliers wrap up at full range and serial multipliers at the other end. Digital serial multipliers anywhere single digits consisting of multiple bits the region is operated at the center of region unit. These multipliers have a direct impact on execution in each speed and space. Duplication is accomplished by including a posting of moved multiplicands in advance with the digits of the number.

Augmentation is an essential rudimentary perform in math activities. Indeed, augmentation-based activities like "multiply and accumulate (MAC)" and spot item region unit among some times utilized calculations concentrated number juggling capacities directly implemented in a few DSP applications like "convolution," FFT, etc.

The DWT algorithm uses constant multiplications. Constant multipliers can be implemented using the shift-add method instead of direct use of hard multipliers to take advantage of the fixed bit pattern of the constant multiplication operand value. In the shift-add method, it multiplies of a variable {x} with the constant value {c} is represented as a sum of shifted version of {x}, where the number of addition operation depends on the number of logics '1' in the 2's complement representation of the constant value. The shifting operation is implemented through hard wiring. Therefore, the combinational logic complexity of constant multiplier based on shift-add method depends on the adder complexity. Different types of number systems have been considered to represent the constant values using less number of logics '1's. CSD is popularly used for representation of constant values [15]. CSD representation uses 33% less logic '1's than those required by 2's complement representation of any signed decimal value. The convolution-based DWT computation involves "inner-product computation." An N-point "inner-product computation" involves N constant multiplications. These constant multiplications are performed in a single stage in parallel and the addition operations are performed in a separate stage. In other words, a set of N constant multiplication operations are performed together in the first stage of computation of N-point inner-product computation. The adder complexity of a set of constant multipliers can be reduced by using "multiple constant multiplication (MCM)" approach. Therefore, CSD representation of constant values and using MCM approach the combinational logic complexity of inner-product design can be reduced substantially than using hard multipliers without compromising on the throughput rate.

A shift-adder-based multiplier has actualized [2]. This work engaged with less rationale assets yet with possibly higher deferral. A SOPOT-based multiplier has designed in Ref. [4]. This work engaged with huge rationale assets and more force utilization. The SOPOT engineering, which includes a flexible shifter and variable shifter, was performed in Ref. [4]. We find that with the flexible shifter, some excess tasks are involved. The primary commitment in this chapter is to supplant one shifter and one multiplexer with an adaptable shifter.

The updated "CSD multiplier" is designed in this exploration work and the equivalent is used to build up a proficient DWT based on lifting. The ASIC findings highlight the competence of the updated multiplier.

2.2   LIFTING SCHEME DISCRETE WAVELET TRANSFORM (DWT)

The development of Bi-symmetrical wavelets for lifting plan was created by Wim Sweldens [5]. The developments are acquired in the spatial space, which is the primary component of this lifting plan. It additionally need not bother with complex numerical counts, which are fundamental in customary techniques. The lifting plan is easiest and able calculation to register wavelet changes. It does not depend on the Fourier transform (Figure 2.1).

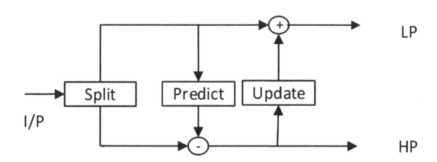

FIGURE 2.1   Block diagram of the lifting-based structure.

1. **Split Step:** The sign is part of the even and odd emphasis of this progression in order to find the most extreme relationship between the nearby pixels to be used during the next planned stage. Each pair of x(n) info tests is divided into odd x(2n + 1) and even x(2n) coefficients.

2. **Predict Step:** The product of the even examples and the predictive factor are given in order to construct the total coefficients (dj) and the output is added to the odd examples in this progression. This result is a high pass screening, as the critical coefficients are considered.

3. **Update Phase:** Here, in this progression, the result of the comprehensive coefficients identified in the last prediction step and the update factors are carried out. Also, examples to obtain the gross coefficients (sj) are added to the results. These contribute to a low

separation of passes. In Figure 2.2, the square diagram of the DWT elevation plane is shown.

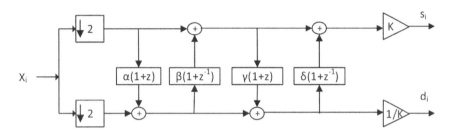

FIGURE 2.2   Block illustration of lifting method DWT.

The lifting coefficients δ, β, α, γ and K are δ ≈ 0.88701370409, γ ≈ 1.7658221510, α ≈ –3.172268684, β ≈ –0.105960237 and K ≈ 0.812893066, respectively.

The complete lifting procedure consists of four steps, in which the input sequence Xi, where i = 0, 1, 2, …. N–1. Here N is the input data range.

- **Splitting Step:**
  Odd part $di(0) = x2i + 1$ (2.1)
  Even part $si(0) = x2i$ (2.2)
- **First Lifting Step:**
  Predictor $di(1) = di(0) + α×(si(0) + si+1(0))$ (2.3)
  Updater $si(1) = si(0)+ β ×(di-1(1) + di(1))$ (2.4)
- **Second Lifting Step:**
  Predictor $di(2) = di(1)+ γ ×(si(1)+ si+1(1))$ (2.5)
  Updater $si(2) = si(1) + δ×(di-1(2) + di(2))$ (2.6)
- **Scaling Step:**
  $di = di(2) /K$ (2.7)
  $si = K ×si(2)$ (2.8)

2.3   SHIFT-ADD MULTIPLIER

The multipliers require a lot of equipment assets in equipment use, and they are difficult for integrated circuit (IC) plans. Here we welcome the step to incorporate activities to boost the duplications, as the wavelet channel coefficients are consistent. That is, first, the coefficients are quantized into a parallel structure and the normal factor is located afterwards.

Alpha = −3.17226868408289)10, (for example, and its paired structure is −11.0010110000011)2. The movements can be proceeded as appeared in Figure 2.3. The comparative technique can be performed for different coefficients. By utilizing the move include activities, the equipment assets are additionally diminished, and the design is more appropriate for equipment usage. The shifting position is represented like this.

$$\alpha = (-1\;1.0\;0\;0\;1\;0\;1\;1\;0\;0\;0\;0\;0)_2$$
$$\uparrow\uparrow\qquad\uparrow\;\uparrow\;\uparrow$$
$$1\;0\qquad -4\;-6\;-7$$

Example:

$$Z = \alpha \times X;$$
$$Z = -X{<}{<}1 - X - X{>}{>}4 - X{>}{>}6 - X{>}{>}7$$

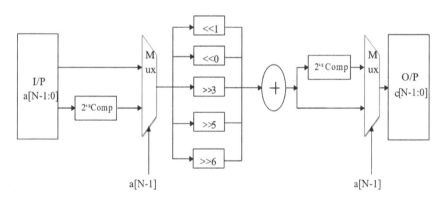

FIGURE 2.3   Proposed architecture for α multiplier using shift adders.

2.4   MULTIPLIER USING CANONICAL SIGNED DIGIT

Accepted marked digit (CSD) is a kind of number portrayal. CSD portrayals have been demonstrated to be helpful in actualizing multipliers with diminished unpredictability, in light of the fact that the expense of increase is an immediate capacity of the quantity of non-zero pieces in the multiplier. CSD portrayals have been demonstrated to be helpful in actualizing multipliers with diminished unpredictability, on the grounds that the expense of duplication is an immediate capacity of the quantity of non-zero bits in the multiplier.

2.4.1 CHARACTERISTICS OF CSD REPRESENTATION

- CSD introduction of a number comprises of numbers 0, 1, and −1.
- The CSD introduction of a number is novel.
- The number of non-zero digits is negligible.
- There cannot be two back-to-back non-zero digits.

Steps for converting a binary number into CSD number:

- Starting from the LSB, check if there are more than one non-zero components (1 or −1).
- Take all non-zero components in addition to the following zero.
- If there is no zero at the left half of the MSB, make one there.
- Add one to the number, i.e., changes the 0 to 1, and all the 1's to 0's and power the furthest right piece to be −1.
- If there are not any more successive non-zero digits, the transformation is finished.

Example for α coefficient:

- Let us take the coefficient α of the lifting structure which is −3.17226868408289.
- Binary representation for the coefficient α is 011.0010110000011.
- There are more than one non-zero digits. Take all the non-zero elements plus the next zero.
- Then add one to the number, i.e., change that 0 to 1 and rightmost bit to −1 and remaining 1 to 0, i.e., 011.001011000010-1.
- Still there are consecutive one's, process repeats, i.e., 011.00110-1000010-1.
- Still there are consecutive one's, process repeats, i.e., 011.010-10-1000010-1.
- Before decimal point there are consecutive ones the same process is applied, i.e., 10-1.010-10-1000010-1.
- The CSD representation for the coefficient α is 10-1.010-10-1000010-1.
- It can be expressed as $-2^2 + 2^0 - 2^{-2} + 2^{-4} + 2^{-6} - 2^{-11} + 2^{-13}$.

The same approach is used for all lifting structure coefficients and values in Table 2.1 are laid down.

2.5   PROPOSED CSD MULTIPLIER

DWT consists of additional and duplicates. The regular DWT engineering with consistent coefficients has increased, which expends more region and power. In DSP calculation, a few incremental tasks are performed. Consistent multipliers are then substituted by appropriate marks [6]. The multipliers are replaced by moving adders that run without a multiplier immediately. Table 2.1 offers a similar technique for both lift structures and efficiency coefficients. The strength of two negative and positive sections is shown in Table 2.1. The positive example and the negative forms talk differently to the left and the right.

TABLE 2.1   CSD Representation of Lifting Coefficients

Coefficient	Coefficient Value	CSD Representation
α	"–3.17226868408289"	"$-2^2 + 2^0 - 2^{-2} + 2^{-4} + 2^{-6} - 2^{-11} + 2^{-13}$"
β	"–0.10596023714940"	"$-2^{-3} + 2^{-6} + 2^{-8} - 2^{-11}$"
γ	"1.76582215102264"	"$2^1 - 2^{-2} + 2^{-6} + 2^{-12} - 2^{-14}$"
δ	"0.88701370409649"	"$2^0 - 2^{-3} + 2^{-6} - 2^{-8} + 2^{-12} + 2^{-14} - 2^{-16}$"
K	"0.81289306611600"	"$2^0 - 2^{-2} + 2^{-4} + 2^{-11} + 2^{-13}$"
1/K	"0.61508705245638"	"$2^{-1} + 2^{-3} - 2^{-7} - 2^{-9} - 2^{-13} - 2^{-15}$"

The engineering authors proposed in Ref. [4] have adaptable shifters and shifter variables. There is no greater intensity of two contrasts (e.g., more extraordinary than seven movements) when you look at Table 2.1. Open shifters are streamlined in these lines, without disrupting the usefulness and accuracy of the multiplier. The modified multiplier technology is shown in Figure 2.4. The planned development is characterized by the SOPOT expression of the switch and the collector. Length of manage registers 8. Information about the purpose the data is not mentioned in the technical presentation. On the basis of this control bit array, the CSD location signal is restricted with the Selk (6) feature, which expands and removes.

If the Selk (6) control bit is coincidentally positive, the manifold expansion starts and the negative manifold fails. Exemplary notation for CSD terminology is limited to Selk (5) and is based on a variable switcher. This control bit moves right and left. If coincidentally Selk (5) is a positive switch of the control bit variable, it commands the left movement, and the negative modulator works to move the right. The main control Selk (2) selects various speed activity bits. Selector control (1: 0), selector move action, bit control panel (7), input/register data selection. Calculations for the lift coefficient α are shown in Table 2.2.

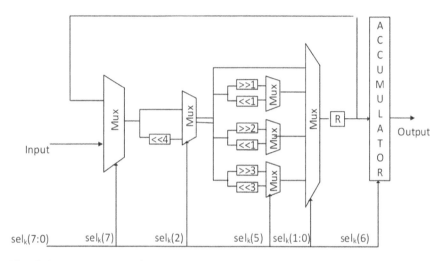

FIGURE 2.4   Proposed architecture for multiplier.

TABLE 2.2   Configurations of the Lifting Coefficient α

CSD Term	Control Register	Configuration
$+2^0$	Sel_0	10000000
-2^2	Sel_1	01000010
-2^{-2}	Sel_2	01100010
$+2^{-4}$	Sel_3	00100010
$+2^{-6}$	Sel_4	00100010
-2^{-11}	Sel_5	01100101
$+2^{-13}$	Sel_6	00100010

2.6   SYNTHESIS RESULTS

The presented effort is planned in "Verilog HDL" and the amalgamation is finished utilizing "Cadence Genus Synthesis" of "90 nm innovation library [7]." The ASIC execution aftereffects of presented, move-adders, and existing SOPOT models are appeared in Table 2.3. Table 2.3 shows the delay, zone, power esteems and abundance territory defer item (EADP), overabundance power postpone item (EPDP) results contrasted with different works for unique piece widths separately. The presented plan has 21.75% and 12.35% less ADP; 21.54% and 22.07% less PDP when contrasted with move snake and SOPOT models, individually for unique bit-widths.

TABLE 2.3   ASIC Implementation Results of Proposed and Existing Designs

Design	Width (n)	Delay (ps)	Area (Cells)	Power (pW)	ADP	PDP	EADP (%)	EPDP (%)
Shi et al. [3]	10	1,380	1,065	236.60	1469700	326509.4	2.34	2.42
	12	1,571	1,672	363.55	2626712	571137.1	32.11	34.62
	16	1,981	2,319	425.40	4393939	842731.3	30.80	27.59
Wu et al. [4]	10	815	1,976	471.96	1610440	384647.4	12.14	20.66
	12	949	2,327	529.40	2208323	502404.4	11.07	18.42
	16	1,222	3,272	687.17	3998384	839727.9	13.84	27.14
Proposed CSD	10	815	1,762	391.13	1436030	318778.3	–	–
	12	949	2,095	447.03	1988155	424238.1	–	–
	16	1,222	2,874	540.47	3512028	660464.1	–	–

2.7 CONCLUSION

This chapter presents a modified CSD architecture with an emphasis on space and energy optimization. Analyzing SOPOT-supported multiplier logical operations, we achieved that this architecture has many redundant processes. The redundant logical operations present in the SOPOT structural design have been removed in the proposed new multiplier by the CSD demonstration. Accomplishment outcomes demonstrated that the presented architecture had 21.75% and 12.35% fewer ADPs. For different bit depths, respectively, matched to the shift adder and SOPOT architectures. The results showed that, relative to the mean shift adder and SOPOT architecture, the architecture was 21.54% lower than the average PDP and 22.07% lower for different bit depth. The results make the tailor-made CSD architecture less space, less power consumption and more effective for the deployment of VLSI hardware.

KEYWORDS

- **canonical signed digit**
- **digital signal processing**
- **discrete wavelet transform**
- **infinite impulse response**
- **lifting**
- **multiplier**

REFERENCES

1. Barsanti, R. J., & Athanason, A., (2013). Signal compression using the discrete wavelet transform and the discrete cosine transform. *IEEE Southeast Conf.*, 1–5.
2. Urriza, Artigas, J. I., Garcia, J. I., Barragan, L. A., Navarro, D., (1998). VLSI architecture for lossless compression of medical images using the discrete wavelet transform. *Proc. Design Automat. Test Eur.*, 196–201.
3. Shi, G. M., Liu, W. F., Zhang, L., Li, F., (2009). An efficient folded architecture for lifting-based discrete wavelet transform. *IEEE Trans. Circuits Syst. II Exp. Briefs, 56*(4), 290–294.
4. Wu, J. F., Ang, A. M. S. Zhang, Z. G., Tsui, K. M., Wu, H. C., Hung, Y. S., & Chan, S. C., (2015). Efficient implementation and design of a new single-channel electrooculography-based human-machine interface system. *IEEE Transactions on Circuits and Systems II: Express Briefs, 62*(2), 179–183.
5. Sweldens, W., (1996). The lifting scheme: A custom-design construction of biorthogonal wavelets. *Appl. Comput. Harmon. Anal., 3*(15), 186–200.
6. Lim, Y. C., & Parker, S. R., (1983). FIR filter design over a discrete power of-two co-efficient space. *IEEE Trans. Acoust., Speech, Signal Process., ASSP-31*, 583–591.
7. Cadence Genus Synthesis Solution, (2015). www/global/en_US/videos/tools/digital_design_signoff/cadence-dac-2015-cadence (accessed on 7 February 2022).
8. Hewlitt, R. M., & Swartzlantler, E. S., (2000). Canonical signed digit representation for FIR digital filters. In: *2000 IEEE Workshop on Signal Processing Systems; SiPS-2000.* Design and Implementation (Cat. No.00TH8528). doi: 10.1109/sips.2000.886740.
9. Jiang, H., Han, J., Qiao, F., & Lombardi, F., (2016). Approximate radix-8 Booth multipliers for low-power and high-performance operation. *IEEE Trans. Comput., 65*(8), 2638–2644.
10. Zhang, W., Jiang, Z., Gao, Z., & Liu, Y., (2012). An efficient VLSI architecture for lifting-based discrete wavelet transform. *IEEE Trans. Circuits Syst. II Express Briefs, 59*(3), 158–162.
11. Darji, R. A., Merchant, S. N., & Chandorkar, A., (2014). Multiplier-less pipeline architecture for lifting-based two-dimensional discrete wavelet transform. *IET Computers & Digital Techniques, 9*(2), 113–123.

12. Mohanty, K., Mahajan, A., & Meher, P. K., (2012). Area and power-efficient architecture for high-throughput implementation of lifting 2-D DWT. *IEEE Trans. Circuits Syst. II Express Briefs, 59*(7), 434–438.

13. Lai, Y. K., Chen, L. F., & Shih, Y. C., (2009). A high-performance and memory-efficient VLSI architecture with parallel scanning method for 2-D lifting-based discrete wavelet transform. *IEEE Trans. Consum. Electron., 55*(2), 400–407.

14. Parhi, K. K., (1999). *VLSI Digital Signal Processing Systems.* John Wiley & Sons.

15. Hartley, R. I., (1996). Subexpression sharing in filters using canonic sign digit multipliers. *IEEE Transactions on Circuits and Systems-II: Analog and Digital Signal Processing, 43*(10), 677–688.

CHAPTER 3

Radix-8 Booth Multiplier in Terms of Power and Area Efficient for Application in Field of 2D DWT Architecture

GUNDUGONTI KISHORE KUMAR[1] and NARAYANAM BALAJI[2]

[1]Department of Electronics and Communication Engineering, V. R. Siddhartha Engineering College, Vijayawada, Andhra Pradesh – 520007, India, E-mail: kishore.chiya@gmail.com

[2]Department of Electronics and Communication Engineering, JNTUK University College of Engineering, Kakinada, Andhra Pradesh – 533003, India

ABSTRACT

This work presents a 16-bit proposed Booth multiplier with a modified 2D DWT architecture. The proposed multiplier gives better performance when compared to existing architecture Canonic Sign Digit (CSD) representation and Booth multiplier in parameters like power, area, and delay. In this proposed method, Booth multiplier, there is no need of generating all the partial products, only the necessary product required by the coefficient is sufficient. By this logic required for generation of partial products gets reduced by which area and power is reduced. In existing 2D DWT architecture with CSD multiplier, it consumes extra clock cycles for the operation, and these extra delay elements are removed which enhances the performance. To show the modified 2D DWT architecture is better than the existing architecture comparison is made with existing 2D DWT architecture with CSD multiplier. It is practicable that for performing the execution, the planned structural design takes fewer digit of clock cycles in comparisons of reported architecture as a result, it found better speed. For the need of comparison with effective results the proposed method, CSD, and Booth multiplier, the synthesis outcomes have been performed using Verilog HDL in Cadence Genus 90 nm tools. There observed an improvement of 29.01% of CSD and 28.22% of existing Booth multiplier with improvement for Excess Area Delay Product (EADP) in excess area than the proposed method and 26.12% and 34.97% for CSD and Booth multiplier in excess than proposed multiplier.

3.1   INTRODUCTION

In mathematical operations for computer application, the multiplication operator plays an important role in this function. It is one of the most common arithmetic operations which play a dominant role in various applications like "fast Fourier transform (FFT)," "discrete wavelet transforms (DWT)," "finite impulse response (FIR)" filters, etc., care implemented for fast moving and low power applications. In DSP system they usually contribute to the time delay and it is a great deal of silicon area in DSP system. As execution time is dominant in most of DSP algorithms, selecting a high-speed multiplier is appropriate.

Multiplier is the arithmetic unit performing multiplication between given numbers and one of the factors deciding the consumption of critical path delay (CPD) and chip area. So, the need of designing a multiplier with less area, low power and fast processing makes a demand for effective implementation of design. Several multiplier designs are developed and among them Booth multiplier is known best for its sign implementation and design of hardware. For minimizing "partial product" arrays in "Booth multipliers" [11], different encoding schemes for Booth algorithm are presented in modified methods. Amongst these, implementation of "radix-8 Booth multiplier" is used for efficient design of hardware and fast multiplication.

Therefore, a whole range of multipliers with very extraordinary zone speed requirements is planned with parallel. Multipliers wrap up at full range and serial multipliers at the other end. Digital "serial multipliers" anywhere single digits consisting of multiple bits the region is operated at the center of region unit. These multipliers have a direct impact on execution in each speed and space. Duplication is accomplished by including a posting of moved multiplicands in advance with the digits of the number. Augmentation is an essential rudimentary perform in math activities. Indeed, augmentation-based activities like "multiply and accumulate (MAC)" and spot item region unit among some of the of times utilized calculations concentrated number juggling capacities directly implemented in a few "digital signal processors (DSP)" applications like "convolution," FFT, etc.

For high-speed processing applications, the demand of developing an efficient solution is of great need and the Booth multiplier is efficient in terms of complexity in area and hardware design. A 12-bit radix-8 Booth multiplier [17] is implemented to reduce overhead complexity by removing extra row in the conventional Booth multiplier. For fast processing speed canonic sign digit [12] builds utilize of add and shift function designed for the execution of the multiplier operation.

DWT is an intelligent, time, and frequency analysis tool with multiple resolutions. The DWT provides higher cryptographic potency, magnitude relationship of compression and image restore value than the usual separate normal feature. As a result, compression formats such as "MPEG-4" and "JPEG 2000" supports the DWT. In signal cryptography [18], the separate moving ridge remodel (DWT) has gained broad unfold acceptance of information compression, information perception, information activity, "geophysics," "meteorology," "audio signal" system, motion pursuit, "machine learning (ML)." In area units, they are particularly appropriate for applications wherever measurability and tolerable degradations are necessary. In contrast to the separate circular function remodel, DWT has higher compression ratios, no obstructive devices and intelligent localization, inherent scaling, and increased versatility in frequency and time domains. Several well-known convolution-based DWT architectures are terribly projected for massive "very large-scale integration (VLSI)" architectures. Later Daubechies and Sweden's consequent the exciting based DWT to cut back advanced computation. The based on lifting DWT has many benefits over convolution-based DWT along with quicker execution, whole number coefficients, absolutely in calculating DWT, a smaller amount hardware quality and isobilateral reverse and forward remodeling. As the lifting scheme offers savings of scaling operations in some multipliers integrated as a square measure executed in a single step, which has advanced CPD than the flip configuration. Therefore, exciting theme gives far good area delay economical configuration than flip formation. The flipping structure projected is a vital DWT design to cut back the essential pathway by removing the multiplier on the trail commencing the participation button to the computing button while not hardware visual projection. The prehistoric information conduit of exciting is changed by inclusion the expect Associate in Nursing update stage to obtain an economical channel design having the essential conduit of one number. The scaling constants of lifting theme square measure excellent inverse of every different whereas the scaling constants of flipping theme are not excellent inverse of every different. In spite of this feature, lifting theme is not therefore standard just similar to the flipping theme to get economical hardware structure for the "2 D-DWT" owing to its longer critical path.

Using high performance reminiscence well-organized structural design for stimulating based architecture 2 D-DWT using parallel scanning method [2]. In first step 4N chronological memory is utilized to save input information and 9/7 filter coefficients. Without utilizing data transposition registers and using data access method in calculating lifting 2D DWT structural design needs 4N +8P bits for on-chip memory [14]. Another lifting data flow [15] by means of parallel computations with pipeline process not including influencing the vital path to reduce data path with symmetrical architecture. A lifting-based DWT [16] for changing serial parallel information and after that send to the column filter for producing 4 sub clusters, after that feed to transposing layer to assure the information flow order required by row filter.

3.2   EXISTING BOOTH MULTIPLIER

Generally unsigned multiplication involves shift and add operations in decimal multiplication. In 1951 Andrew Donald Booth proposed an algorithm which gives access to both +ve and −ve products. To calculate multiplication, result the partial products involve additions and subtractions. The subtraction operation is performed likewise addition operation but it involves 2's complement with sign representation. To increase the speed, booth used desk calculators to shift and add at a faster rate and created the algorithm.

One of the successful algorithms for the signed multiplication of numbers is the booth algorithm, which, in view of the binary p-bit numbers, may be represented by asp/2-digit radix-4 numbers, p/3-digit radix-8 numbers, and so on. The algorithm of modified Radix-8 reduces to p/3 partial products. A multiplier for Radix-8 booth coding uses 4-bit encoding to generate a third of the partial product number. Radix-8 The same algorithm as Radix-4 is applied to the booth recoding method but we take now quartets of bits rather than three. Each quartet has a signed numerical code. Radix-8 decreases the "partial products" to p/3 and n is the multiplier bits. Thus, the partial product summation makes a benefit in time.

3.2.1   PARTIAL PRODUCTS STRUCTURE

In 2's supplement arithmetic, the partial result of booth coded multiplication is represented. The PPS unit produces partial product values of size $(n+2)$ bits corresponding to a particular Booth encoded digit $\{d_i\}$. These $(n+2)$ partial product bits form a row in the partial product row. For a $(n \times n)$ multiplication, the PPS unit produces 'w' fractional product value consequent to 'w' booth encoded digit and these w partial product values forms w rows in the partial product array (PPA). Every fractional product string is missing shift by 3-bits position with esteem to its preceding row. Due to 2's complement representation, symbol expansion of incomplete result rows is essential up to "$(2n-1)$" bit arrangement of "$2n$" bit product. To find the right multiplication result. Addition of sign bit up to "$(2n-1)$" bit point increases the adder unit complication considerably. By inserting the suitable guard bit in every fractional product line to the conservatory of sign bit up to $(2n-1)$ bit location can be avoided. Figure 3.1 shows the conventional fractional product selection of radix-8 Booth multiplication for $n=16$ where guard bits are used for sign extension. Figure 3.2 shows the existing Booth multipliers PPA using dot representation where the extra row (n5) in the conventional Booth multiplier is compensated by absorbing the control bit (n_1) and the modified bits are represented as modified bits ($m_{1,1}$, $m_{1,0}$), carry bit (r_1). By this, the extra row in the conventional method is reduced to partial product row thereby reducing area.

31	30	29	28	27	26	25	24	23	22	21	20	19	18	17	16	15	14	13	12	11	10	9	8	7	6	5	4	3	2	1	0	n
											~(p0,15	p0,15	p0,15	p0,15	p0,15	p0,15	p0,14	p0,13	p0,12	p0,11	p0,10	p0,9	p0,8	p0,7	p0,6	p0,5	p0,4	p0,3	p0,2	p0,1	p0,0	n0
										1	~(p1,15	p1,15	p1,15	p1,14	p1,13	p1,12	p1,11	p1,10	p1,9	p1,8	p1,7	p1,6	p1,5	p1,4	p1,3	p1,2	p1,1	p1,0				n1
							1	1	~(p2,15	p2,15	p2,14	p2,13	p2,12	p2,11	p2,10	p2,9	p2,8	p2,7	p2,6	p2,5	p2,4	p2,3	p2,2	p2,1	p2,0							n2
				1	1	~(p3,15	p3,15	p3,14	p3,13	p3,12	p3,11	p3,10	p3,9	p3,8	p3,7	p3,6	p3,5	p3,4	p3,3	p3,2	p3,1	p3,0										n3
		1	~(p4,15	p4,15	p4,14	p4,13	p4,12	p4,11	p4,10	p4,9	p4,8	p4,7	p4,6	p4,5	p4,4	p4,3	p4,2	p4,1	p4,0													n4
1	p5,15	p5,14	p5,13	p5,12	p5,11	p5,10	p5,9	p5,8	p5,7	p5,6	p5,5	p5,4	p5,3	p5,2	p5,1	p5,0																n5
z31	z30	z29	z28	z27	z26	z25	z24	z23	z22	z21	z20	z19	z18	z17	z16	z15	z14	z13	z12	z11	z10	z9	z8	z7	z6	z5	z4	z3	z2	z1	z0	

FIGURE 3.1   Structure of fractional product array for conventional booth multiplier.

31	30	29	28	27	26	25	24	23	22	21	20	19	18	17	16	15	14	13	12	11	10	9	8	7	6	5	4	3	2	1	0
											t3	t2	t1	t0	p0,15	p0,15	p0,14	p0,13	p0,12	p0,11	p0,10	p0,9	p0,8	p0,7	p0,6	p0,5	p0,4	p0,3	p0,2	m1	m0
										1	~(p1,16	p1,15	p1,15	p1,14	p1,13	p1,12	p1,11	p1,10	p1,9	p1,8	p1,7	p1,6	p1,5	p1,4	p1,3	p1,2	m11	m10	r0		
							1	~(p2,16	p2,16	p2,15	p2,14	p2,13	p2,12	p2,11	p2,10	p2,9	p2,8	p2,7	p2,6	p2,5	p2,4	p2,3	p2,2	m21	m20	r1					
				1	~(p3,16	p3,16	p3,15	p3,14	p3,13	p3,12	p3,11	p3,10	p3,9	p3,8	p3,7	p3,6	p3,5	p3,4	p3,3	p3,2	m31	m30	r2								
	1	~(p4,16	p4,16	p4,15	p4,14	p4,13	p4,12	p4,11	p4,10	p4,9	p4,8	p4,7	p4,6	p4,5	p4,4	p4,3	p4,2	m41	m40	r3											
1	p5,15	p5,14	p5,13	p5,12	p5,11	p5,10	p5,9	p5,8	p5,7	p5,6	p5,5	p5,4	p5,3	p5,2	m51	m50	r4														
z31	z30	z29	z28	z27	z26	z25	z24	z23	z22	z21	z20	z19	z18	z17	z16	z15	z14	z13	z12	z11	z10	z9	z8	z7	z6	z5	z4	z3	z2	z1	z0

FIGURE 3.2   Existing radix-8 16-bit booth multiplier partial product structure.

3.2.2 BOOTH ENCODER (BE) AND SELECTOR UNITS

The basic building block of radix-8 Booth multiplier is revealed in Figure 3.3 for 12-bit structure. The biased creation unit generate the unfair product set {4A, 3A, 2A, A} structure contribution operand 'A' with two shifters and one adder. The "booth encoder unit (BEU)" includes of 'W' "Booth encoder (BEs)." Where every produced the four direct signal BE.

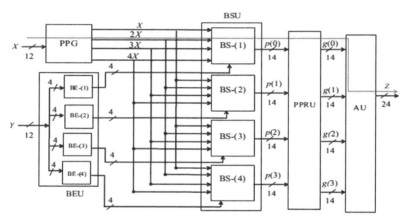

FIGURE 3.3   Block diagram of 12-bit booth multiplier for radix-8.
Source: Reprinted with permission from Ref. [17]. © 2017 Springer Nature.

Suppose 2 m-bit numbers P and Q. where Q is multiplier and P is multiplicand. The multiplier and multiplicand are representing in two's accompaniment form as follows:

$$P = -p_{m-1} 2^{m-1} + \sum_{i=0}^{m-2} a_i 2^i \tag{3.1}$$

$$Q = -q_{m-1} 2^{m-1} + \sum_{i=0}^{m-2} b_i 2^i \tag{3.2}$$

where; p_{m-1}, q_{m-1} of P and Q are sign bits, i.e., most significant bit.

The multiplier algorithm distils the 'm' bit multiplier into $v = [m/3]$ group of three bit each by considering Radix-8 recoding. Every collection consists of one overlapping bit and three bits for 4-bit digit. When considering i_{th} digit $f_i = \{b_{3i+2},\ b_{3i+1},\ b_{3i}\}$, b_{3i-1} is the overlap bit belong to $(i-1)^{th}$ digit f_{i-1} and f_{i-1} = 0, the decimal equivalent of f_i is getting by the relation:

$$f_i = -4b_{3i+2} + 2b_{3i+1} + 3b_{3i} + b_{3i-1} \tag{3.3}$$

Putting Eqn. (3.3) in Eqn. (3.2), we have:

$$B = \sum_{i=0}^{v-1} f_i 2^{3i} \tag{3.4}$$

Substituting Eqns. (3.3) and (3.4) in product of A and B we get that:

$$Z = A \sum_{i=0}^{v-1} \left(-4b_{3i+2} + 2b_{3i+1} + b_{3i} + b_{3i-1}\right) 2^{3i} \tag{3.5}$$

For radix-8 Booth encoding, k=3 is assumed for grouping of multiplier bits. Every group of one overlapping bit and three bits form one digit of size 4 bits, where the ith digit "$d_i = \{b_{3i+2}\ b_{3i+1}\ b_{3i}\ b_{3i-1}\},$" b_{3i-1} is the over-lapping bit belong to $(i-1)^{th}$ digit d_{i-1} and it is zero for $i = 0$ (Figure 3.4).

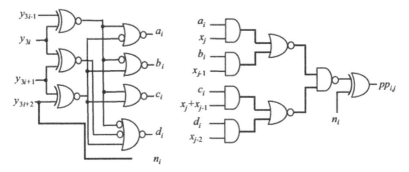

FIGURE 3.4 Inside construction of booth selector (BS) and booth encoder (BE).

The required 16 "partial product" value of "radix-8 Booth encoded multiplication" are obtained from positive partial product values "{0, P, P, 2P, 2P, 3P, 3P, 4P}" and the negative partial product values {−P, −P, −2P, −2P, −3P, −3P, −4P} are obtained from the positive partial product set using a sign control signal (b_{3i+2}). By using the following logic expressions all the required bits in PPA are generated:

$$n_{i0} = ppv_{i0} \oplus mi \tag{3.6}$$

$$r_i = ppv_{i,1}ppv_{i,0}m_i,\ 0 \le i \le \frac{m}{3} - 2 \tag{3.7}$$

$$n_{i1} = ppv_{i,1}\left(\sim ppv_{i,0}\right) + ppv_{i,0}\left(\sim m_i\right) + \left(\sim ppv_{i,1}\right)ppv_{i,0}m_i \qquad (3.8)$$

$$n_{32} = ppv_{i,3}\sim ppv_{i,2} + ppv_{i,2}\sim ppv_{i,0} + ppv_{i,2}\sim m_3 + \left(\sim ppv_{i,2}\right)ppv_{i,1}ppv_{i,0}m_3 \quad (3.9)$$

$$n_{33} = ppv_{i,3}\sim ppv_{i,2} + ppv_{i,2}\sim ppv_{i,0} + ppv_{i,3}m \sim_3 + \left(\sim ppv_{i,3}\right)ppv_{i,2}ppv_{i,1}m_3 \quad (3.10)$$

$$r_3 = ppv_{3,1}ppv_{3,2}ppv_{3,0}m_3 \qquad (3.11)$$

$$q_0 = ppv_{0,13} \oplus r_3 \qquad (3.12)$$

$$q_3 = \sim ppv_{0,13}\left(\sim r_3\right) \qquad (3.13)$$

$$\text{``}p_i = \sim((b_{3i-1}\odot b_{3i})\sim(b_{3i+1}\odot b_{3i+2}))\text{''} \qquad (3.14)$$

$$\text{``}g_i = \sim(\sim(b_{3i-1}\odot b_{3i}))(b_{3i}\odot b_{3i+1})\text{''} \qquad (3.15)$$

$$\text{``}s_i = \sim((b_{3i-1}\odot b_{3i})\,(b_{3i+1}\odot b_{3i+2}))\text{''} \qquad (3.16)$$

$$\text{``}t_i = \sim(\sim(b_{3i-1}\odot b_{3i})\sim(b_{3i+1}\odot b_{3i+2})\,(b_{3i+1}\odot b_{3i+2}))\text{''} \qquad (3.17)$$

The Eqns. (3.6)–(3.17) are the expressions required for encoded bits in Figure 3.2 which complete the partial product structure.

3.3   PROPOSED BOOTH MULTIPLIER

In the existing Booth multiplier area gets increased due to making of fractional products. Every one of the fractional products is generated without the necessity of partial products. So, the need of designing an efficient Radix-8 Booth multiplier is required.

3.3.1   METHOD OF IMPLEMENTATION

Let M and N be two signed numbers of p-bit size, where M is the multiplicand and N is the multiplier which are represented in 2's complement form with using Booth algorithm as multiplication method is shown below:

$$M = -m_{p-1}2^{p-1} + \sum_{j=0}^{p-2}m_j 2^j \qquad (3.18)$$

$$N = -n_{p-1}2^{p-1} + \sum_{j=0}^{p-2} n_j 2^j \tag{3.19}$$

where; m_{p-1}, n_{p-1} of M and N represent extension of sign bits.

Using radix-8 encoding scheme Booth multiplier distils the p-bit multiplier into $k = \lceil p/3 \rceil$ clusters of 3-bits. Grouping of bits is done from right hand side with appending of '0' at the LSB position and grouping of three bits with a fourth bit as overlapping bit is taken. The jth digit a_j can be defined as:

$$a_j = \{n_{3j+2}\ n_{3j+1}\ n_{3j}\ n_{3j-1}\} \tag{3.20}$$

Where the overlapping bit is represented in $(j-1)^{th}$ position as n_{3j-1}. The evaluation of a_j is gained with help of this equation:

$$a_j = -4n_{3j+2} + 2n_{3j+1} + n_{3j} + n_{3j-1} \tag{3.21}$$

Substituting (3.21) in (3.19) we get:

$$N = \sum_{j=0}^{k-1} a_j 2^{3j} \tag{3.22}$$

The multiplication of M and N is obtained by substituting (3.21) and (3.22) in the product output as:

$$N = \sum_{j=0}^{k-1} (-4n_{3j+2} + 2n_{3j+1} + n_{3j} + n_{3j-1})2^{3j} \tag{3.23}$$

Starting number sequence {−4, −3, −2, −1, 0, 1, 2, 3, 4} and a_j selects the decimal significance, and when M is multiplied by a_j It produces a product expression (M.a$_j$) which outcomes in getting essential partial results of radix8 Booth multiplier are {−4M, −3M, −2M, −M, 0, M, 2M, 3M, 4M}. Each of the partial product operation is obtained through shifting and adding the encoded term. The positive partial product {M} is obtained by placing the same partial product of p-bit size and the {2M, 4M} partial product terms are obtained by shifting M to left by one bit and two bits. The product term {3M} is obtained by adding M and 2M partial products. For obtaining the negative partial products {−4M, −3M, −2M, −M} 2's harmonize of respective positive terms can be done in two steps. First the 1's complement of {M, 2M, 3M, 4M} has found and in subsequently stage to get the negative product expressions the sign bit (s$_j$) is added to the respective product term. Generated partial product terms {M, 2M, 3M, 4M} are produced from partial product generator which has to be selected through booth selector (BS) either in normal or complemented form. BS uses $\{e_j, f_j, g_j, h_j\}$ as control signals to select one value from the set $\{M, 2M, 3M, 4M\}$ and the sign bit (s_j) is to generate 1's complement of partial product. The control signal of respective bit becomes enable:

$e_j = 1$ (enabled) when "partial product" term $P = \{M, -M\}$;

$f_j = 1$ (enabled) when "partial product" term $P = \{2M, -2M\}$;

$g_j = 1$ (enabled) when "partial product" term $P = \{3M, -3M\}$;

$h_j = 1$ (enabled) when "partial product" term $P = \{4M, -4M\}$;

$s_j = 1$ (enabled) when "partial product" term $P = \{-4M, -3M, -2M, -M\}$;

"Radix-8 Booth encoding" scheme to generate "partial product" bits pp_{ji} for $0 \le j \le k-1$, $0 \le i \le p-1$ is shown in Table 3.1.

TABLE 3.1   Radix-8 Booth Encoder Partial Product Values

$n_{3j+2}n_{3j+1}n_{3j}n_{3j-1}$	Value	e_j	f_j	g_j	h_j	s_j	pp_{ji}
0000	0	0	0	0	0	0	0
0001	+M	1	0	0	0	0	m_j
0010	+M	1	0	0	0	0	m_j
0011	+2M	0	1	0	0	0	m_{j-1}
0100	+2M	0	1	0	0	0	m_{j-1}
0101	+3M	0	0	1	0	0	$m_j + m_{j-1}$
0110	+3M	0	0	1	0	0	$m_j + m_{j-1}$
0111	+4M	0	0	0	1	0	m_{j-2}
1000	−4M	0	0	0	1	1	$\overline{m_{j-2}}$
1001	−3M	0	0	1	0	1	$\overline{m_j + m_{j-1}}$
1010	−3M	0	0	1	0	1	$\overline{m_j + m_{j-1}}$
1011	−2M	0	1	0	0	1	$\overline{m_{j-1}}$
1100	−2M	0	1	0	0	1	$\overline{m_{j-1}}$
1101	−M	1	0	0	0	1	$\overline{m_j}$
1110	−M	1	0	0	0	1	$\overline{m_j}$
1111	0	0	0	0	0	0	0

Note: m_j represent the same partial product to be retained where e_j gets activated; m_{j-1} represents the one bit be shifted to left where f_j gets activated; $m_j + m_{j-1}$ representing the $\{M\}$ and $\{2M\}$ to be added where g_j gets activated; m_{j-2} denotes shifting of two bits to left where h_j gets activated; "$\{m_j, m_{j-1}, m_j + m_{j-1}, m_{j-2}\}$" denotes the complemented forms of respective terms where w.r.t. control bits $\{e_j, f_j, g_j, h_j\}$, sign bit s_j gets activated.

Let us consider an example (for example, M = 1/α = –0.630464) which explains how it is encoded and selection of particular product term is done is explained. First consider the coefficient with bit size and note the partial products with sign consideration. In this example, a 16-bit size coefficient is considered, as the coefficient is negative term, it has to be converted into 2's complement (1's complement + 1) form. If the coefficient is positive then directly it can be noted (Table 3.2).

TABLE 3.2 Partial Product for M = 1/α Coefficient of 16-Bit Width

Coefficient	15	14	13	12	11	10	9	8	7	6	5	4	3	2	1	0
A = 0.630464	0	0	0	0	1	0	1	0	0	0	0	1	0	1	1	0
~A	1	1	1	1	0	1	0	1	1	1	1	0	1	0	0	1
																1
A = –0.630464	1	1	1	1	0	1	0	1	1	1	1	0	1	0	1	0

The input (for example, N=99) which is a variable term has to be encoded with appending '0' at the LSB position (shown in black color) and encoding each part with 4 bits. For each encoded term it has to be selected from the selector unit to recognize the respective product term which is shown in Table 3.3.

TABLE 3.3 Booth Encoding Scheme of Radix-8 16-Bit for N = 99

0						+2						+3			
0	0	0	0	0	0	0	0	0	0	0	1	1	0	0	0	1	1	0
	0					0						–4			

3.3.2 PARTIAL PRODUCT ARRAY (PPA) STRUCTURE

The structure of 16-bit Radix-8 PPA and Booth multiplier depends. When constant value is multiplied by the multiplier, it can be referred to as 'fixed width coefficient.' Suppose the constant multiplier has $p \, X \, p$ bit size then the resultant output has $2p$ output bit size. The number of partial product rows obtained are based on k = [p/3] of p bit size. For 16-bit partial product size the k comes to a value of 16/3 = 6 rows, so on a whole in 16-bit, it consists of six rows. Each partial product row w. r. t previous row it is shifted to left by three positions. To enhance correct multiplication output the extension of sign bits are required till $(2p-1)$ bit positions but which increases the complexity of the adder. To avoid extension of sign bits up to $(2p-1)$ bit positions, guard bits are introduced which is shown in Figure 3.5. As discussed above, for each row control bit (s_j) is added to convert "1's complement" number to "2's complement." To the next row at the starting position, s_j is added to determine whether that partial product row is positive or negative is shown in Figure 3.5. By adding s_j to the PPA the product rows get increased by one extra row, thereby which impacts extra adder stage complexity. So, to compensate the control bit carry bit (r_j) is added as in Figure 3.6. Mohanty et al. [17] has given a structure by modifying the regular PPA to eliminate the extra row. When we are adding each column of the row, there needs an extra adder, and this can be minimized by compensating with carry bits (q_j) (Figure 3.5).

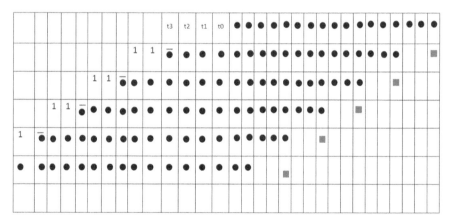

FIGURE 3.5   Conventional "16-bit radix-8 booth multiplier partial product" terms where ● represents the sign bit; ■ are the partial products; q0, q, q2, q3 are the guard bits represented in dot representation.

In each row by absorbing the sign bit s_j, the two LSB positions ($pp_{j,1}$, $pp_{j,0}$) are replaced with modified bits ($q_{j,1}$, $q_{j,0}$) of $(j+1)^{th}$ row in generating the carry bit r_j. In generating the carry bit (r_{k-1}) the control bit (s_j) is placed at the right of $(j+2)^{th}$ LSB row of PPA. To produce the last carry bit (r_{k-1}) with sign addition bits of primary row they are customized into (t3, t2, t1, t0) LSB's of kth row PPA ($pp_{k-1,1}$, $pp_{k-1,0}$) with modified bits ($q_{k-1,1}$, $q_{k-1,0}$). The expressions for modified bits ($q_{j,1}$, $q_{j,0}$), carry bit (r_{k-1}) produced are shown below:

$$q_{j0} = pp_{j0}{}^{\wedge}S_j \qquad (3.24)$$

$$q_{j1} = pp_{j1}{}^{\wedge}(pp_{j0}\&s_{j1}) \qquad (3.25)$$

$$r_j = pp_{j^\prime}pp_{j,0}S_j \; 0<j<\frac{p}{3}-2 \qquad (3.26)$$

$$t_0 = pp_{0,15}{}^{\wedge}r_5; \qquad (3.27)$$

$$t_3 = \overline{p_{0,15}r_5} \qquad (3.28)$$

$$t1, t2 = \overline{q3} \qquad (3.29)$$

In the proposed method, generation of all partial products are not required so that for selection and generation, a smaller number of elements is used. So that the time required for each element also gets decreased due to which improvement in the design is improved. By determining the scaling terms the generation of product terms can be known and no need of producing other terms. The partial product structure for $1/\alpha$ coefficient is shown in Figure 3.7.

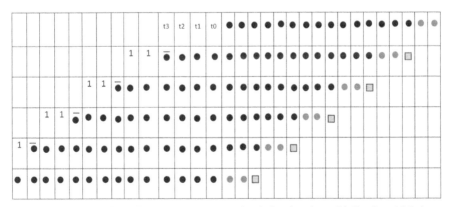

FIGURE 3.6 Partial product array structure of existing booth multiplier in which it ● represents the modified bits; ☐ represent the carry bits using dot representation.

3.3.3   *STEPS INVOLVED*

Basically, in this era of technology, every application requires needs parameters which occupies a small area with low power consumption and increase in speed. So, the need of designing multiplier in this chapter focus on low area and power and this is achieved by reducing partial products. To reduce the area only required partial products are generated which can be known by scaling the coefficient, which also improves the performance. To eliminate infinite increase in bit width in various algorithms of VLSI implementations fixed width multipliers are used. The output of fixed width multiplier is same as of input operand. A full-width multiplier design involves truncation of 2p bit size output to p bit size output by either post truncation or truncated multiplier. In this proposed multiplier for a p-bit size input, the length of the obtained output is 2p which is post-truncated to p bit size of full width multiplier. On doing truncation for a multiplier the required number of logic resources also becomes less than the fixed-width multiplier and by taking care on the carry bit there would be compensated in minimum truncation error.

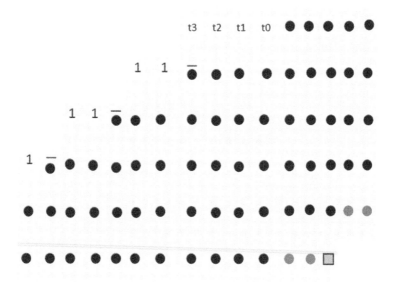

FIGURE 3.7   Proposed partial product array structure in which ● represents the modified bits; ☐ represents the carry bits using dot representation.

In the proposed multiplier all the partial products are not required to generate, only necessary partial products are sufficient to complete a multiplier. The generation of partial products may vary depending on the multiplier coefficient and for making multiplier design in low area and power, doing the truncation in left hand side of PPA structure. Because when considering the PPA in Figure 3.5 on right hand side generation of partial products are there and on left hand side extended bits are considered in which no need of extra logics, so when truncating if we are post-truncated the bits to left side a minimization in area and power is achieved. The scaling done to each co-efficient is not uniform scaling as we try to reduce the terms in generation and by this, an error rate of +/– is introduced which is acceptable in image processing applications.

Consider a constant value (M) represented in 16-bit number which is to be multiplied with variable value (N) represented in 16-bit number. When M and N are multiplied, the resultant output is of 32-bit length. This 32-bit output value is post truncated to 16-bit length with truncation of both higher and lower order bits as in Figure 3.6. The truncation depends on scaling coefficient which varies from coefficient to coefficient. Let M be $(1/\alpha)$ = –0.630464 used in 2D DWT architecture and the steps involved are:

> **Step 1:** For the coefficient (M) mention the scaling range.
> **Step 2:** From the booth encoder list the required "partial product" terms that should be generated.
> **Step 3:** Encoding and selection is done in booth encoder and selector and the respective each column "partial product" terms are added using "carry save adder."
> **Step 4:** The post-truncation of the output depends on scaling range.

The above steps are followed in an example, shown in the next section in that the output is taken from 12th bit to 27th bit as shown in Figure 3.7.

3.3.4   THEORETICAL ANALYSIS SHOWING WITH AN EXAMPLE

Taking an example for the proposed Booth multiplier with M = –0.630464 as multiplicand and N=99 as multiplier. When M is multiplied to N how the theoretical analysis is done is shown in Figure 3.8 with generation of "partial products." As encoding is done to M and selection is taken from N shown in Table 3.3. After obtaining all the partial products, each column is added using a "carry save adder" to obtain the output. The scaling range of M is 12 and in the resultant, the output is taken from 12th bit to 27th bit. Generation of bits between 12th bit and 27th bit is sufficient to obtain the resultant output with an error rate of +/–1%. In this example when M = –0.630464 is multiplied with N=99 the calculated output is –62.415 and the theoretically obtained output is –62 which is accepted within the error rate of +/–1%. If the coefficient is a signed number, the output is also represented in signed number, so it converted into 1's complement number to represent the output.

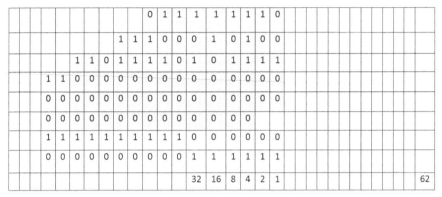

FIGURE 3.8 Example showing the proposed multiplier with M= –0.630464 and N = 99 for the coefficient (1/α = –0.630464).
Source: Reprinted from Ref. [12]. Open access. © 2017 John Wiley.

The above example shows the one coefficient used in 2D DWT and for different coefficients the scaling range varies. The truncation used for coefficients in 2D DWT is non-uniform scaling as in this multiplier it tries to reduce the partial product terms by doing scaling at the generation part so that circuits needed to generate them gets reduced. In 2D DWT architecture, the various coefficients used, value, Precision values, etc., are planned in Table 3.4. The exceeding Table 3.4 shows how the scale be supposed to be done for a variety of coefficient. By use of truncation method, the requirement for quantity of adder units. The creation of incomplete term gets to be reduced, and no difficulty of enchanting the productivity from the accuracy bit is simple.

TABLE 3.4 Precision Values of DWT Coefficients

Coefficient	Decimal Value $(X2^{-precision})$	Precision	Value	Binary Value
(1/α)	–2582	12-bit	–0.630464	1111010111101010
(1/α × β)	6093	9-bit	11.900004	0001011111001100
((1/δ × γ) + 1)	7278	11-bit	3.553775	0001110001101110
(1/β × γ)	–5473	8-bit	–21.37815	111010101010000
((1/α × β) + 1)	6605	9-bit	12.900004	0001100111001100

3.4 IMPLEMENTATION OF 2D DWT USING PROPOSED BOOTH MULTIPLIER USING DWT

The VLSI implementation of 2D DWT of image encoders with flipping based structure is addressed. For applications like image processing, the demand for DWT is also increasing day-by-day. It has advantages over other transform like discrete cosine transform (DCT) in representation of an image in coarse form which is subset of coefficients of filter and the inverse transform can be computed without computing. One of the applications is it can be used for internet applications using progressive image transmission (PIT). Based on different areas of interest, DWT permits to encode the image in different factors with compression factors, levels of implementation and type of architectures.

Wavelets exchange the image into a movement of wavelet that can be stored up more successfully than pixel squares, so "DWT" models have picked up its significance in applications where adaptability and okay corruption is fundamental in wavelet coding plans. At the point when time and recurrence spaces are found, the middle value of for entire length of the sign gives data about the "DWT." "2D-DWT" is utilized widely in numerous fields of building and clinical applications, for example, in "biometrics," picture investigation and imaging applications, for example, "JPEG 2000" and so forth.

3.4.1 IMPLEMENTATION OF EXISTING 2D DWT ARCHITECTURE

In the existing 2D DWT architecture it aims to develop without off-chip RAM with high-throughput and makes use of canonic sign digit multiplier in replacement to traditional multipliers to gain high hardware efficiency. For an image size of N X N, a three-level 2D DWT is implemented which requires 66 subtractors, 123 adders, 7.5 words of temporal memory, 3N bytes of input RAM and 167 registers. This 3-level 2D DWT is 14.1% lower than current architectures, which boosts transistor delay products. This architecture takes almost a total pause on the "critical path."

One-level "2D-DWT": The column filter, the unit and the line-filter shown in Figure 3.9. The column filter is a structure with three inputs in which two outputs are provided as low and high signals. The two output signals from the column filter are passed through the transposing unit followed by the row filter which gives two signals as output.

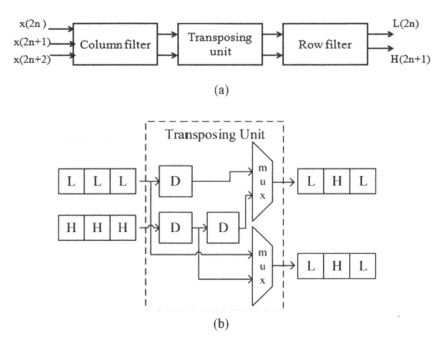

(a)

(b)

FIGURE 3.9 (a) One-level 2-D DWT architecture; (b) transposing unit.
Source: Reprinted from Ref. [12]. Open access. © 2017 John Wiley.

The structure of column filter and row filter is shown in Figure 3.10, which makes use of multipliers, adders, and registers for implementing the one-level DWT. The multiplier uses a Canonic sign Digit representation for minimum CPD and fast speed of operation. For the adder circuit using the ripple carry adder as it involves shorter delay and the delays in the circuit is introduced by using D flip-flop. When three inputs are given to the column filter it gives "low band (L) and high band (H)" as output. When the L and H band signals are applied to transposing unit, it gives two outputs (L and H). When applied to row filter it gives output signals as HL/LL as one output and HH/LH as the other output. In the existing 2D DWT architecture RAM to make it useful without off-chip RAM. Generally, the hardware resources consist of two parts: one is RAM requirement and the other is the computational core. For RAM requirement, it consists of input RAM and temporal RAM. The input RAM stores the data and size of the input depending on external bandwidth and scanning method. Temporal method depends on the architecture and is used to store the calculations obtained in the result.

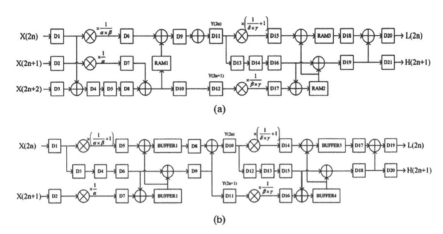

(a)

(b)

FIGURE 3.10 (a) Column filter; (b) row filter.

In the column and row filter the coefficients used for multiplier are represented in Canonic Sign Digit format. Taking an input value X multiplying with the coefficient, then the output will be constant multiplied by the input value. The structures of one by alpha coefficient value is shown in Figure 3.11. Each constant value has a different structure because by varying each value the structure becomes different by adding shift and adder units. In the below structures the D represents the delay in the circuit and we use D flip flop as the delay element. Ripple carry adder is used as the adder and << denotes the left shift operation for the input to that register and >> denotes the right-hand shift operation for the input to that register.

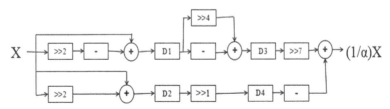

FIGURE 3.11 CSD multiplier structure for $1/\alpha$ coefficient.
Source: Reprinted from Ref. [12]. Open access. © 2017 John Wiley.

Considering an example for $1/\alpha = -0.630464$ coefficient with CSD multiplier design and assuming the input number as 0×0054 (hex decimal number system). Table 3.5 shows how the input value is processed within how many clock cycles is described.

TABLE 3.5 Example Showing CSD Multiplier for $1/\alpha = -0.630464$

Clock Cycle	X	D1	D2	D3	D4	Result
1	0×0054	–	–	–	–	–
2	–	0×003F	0×0069	–	–	–
3	–	–	–	0×FFDC	0×0034	–
4	–	–	–	–	–	0×FFCB

Source: Reprinted from Ref. [12]. Open access. © 2017 John Wiley.

The CSD multiplier consumes four clock cycles for the operation. Considering the structure of $1/\alpha$ coefficient given an input value X = 0×0054 and this X passes through the series of operations like >>2, >>4, >>1, >>7, the ripple carry adder in the adder circuit and 2's complement is used in place of subtractor block. The D represents the D flip-flop in the structure which makes use of clock for the operation. Initially when the reset is 0 the D registers are set to 0 and when reset is 1 the register output is same as input. In the first clock cycle all the registers (D1–D4) have no value because D flip-flop takes one cycle to get the output and in second clock cycle after series of operation the D1 and D2 register gets a value of 0×003F and 0×0069 because of parallel connection in structure. Next in third clock cycle the D3 and D4 gets a value of 0×FFDC and 0×0034 and in fourth clock cycle the resultant output is obtained with a value of 0×FFCB. When multiplying the X = 84(0×0054) with –0.630464 the theoretical output is –52.958 and the observed output is –52 (0×FFCB) which is acceptable within the +/–1% error rate.

When the same $1/\alpha$ coefficient is done with the proposed Booth multiplier, then it takes only one clock cycle for the operation which helps in reducing the latency in the structure. When the same input value X=84 is applied to the multiplier it gives an output value –53 (0×FFCA) shown in Table 3.6. Compared to existing Booth multiplier, the numbers of partial products are reduced, and generation of product terms are also reduced, which improves hardware efficiency.

TABLE 3.6 Example Showing Proposed Multiplier for $1/\alpha = -0.630464$

Clock Cycle	X	Result
1	0×0054	0×FFCA

3.4.2 IMPLEMENTING 1D DWT ARCHITECTURES USING PROPOSED BOOTH MULTIPLIER

In the 2D DWT, because of a reduction in the number of partial product terms, the architecture replacing the CSD multiplier with the proposed multiplier decreases the area and power. The 1D DWT occupies 12 clock cycles in the current architecture (i.e., CSD implementation), and the clock cycles are decreased in the suggested architecture by evacuating the additional 4 register units since it occupies 8 clock cycles. By increasing the rpm, this reduction in clock cycles affects efficiency.

The optimized column and row filter for the 1D DWT are shown in Figures 3.12 and 3.13 where the 4 extra D flip-flops are omitted because of the removal of the additional delay applied to the circuit. The 2D DWT architecture consists of a column filter, a row filter, and a transposition unit. 1D column filter is processed along column where it is a three input and gives output as two-output (low (L) and high (H)) sub-bands. These two L and H sub-bands are passed through transposing unit where it gives two sub-bands. The obtained sub-bands (L and H) are passed through row filter where it is processed along row wise and gives HL/LL as one output and HH/LH as other output. On comparing the existing 2D DWT architecture and optimized 2D DWT architecture, we can observe a reduction in a number of clock cycles by removing 4 delay elements in each row and column filter. By this speed is improved and hence it increases the performance of the proposed architecture.

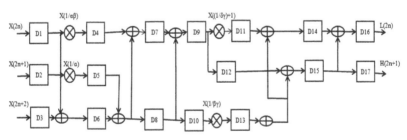

FIGURE 3.12   Modified column filter.
Source: Reprinted from Ref. [12]. Open access. © 2017 John Wiley.

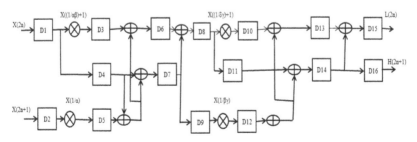

FIGURE 3.13   Modified row filter.
Source: Reprinted from Ref. [12]. Open access. © 2017 John Wiley.

The utilization of hardware components in "2D-DWT" is shown in Table 3.7. The elements used in 2D DWT are ripple carry adder is used as adder circuit, D flip-flops and in case of multipliers CSD, Booth multipliers are

used. In total, the total number of components comprises of "AND/OR/NOR/
NAND," "XOR/XNOR" gates, latency, and registers required for 2D-DWT
implementation for the listed multiplier architectures. Compared to CSD
implementation, the proposed multiplier needs less number of registers which
is an advantage to various image processing applications for occupying less
area. Normally in CSD representation for execution of output, the storage
element ("D flip-flop") requires two clock cycles, and in column filter to
compensate the effect extra two registers are added using "CSD" implementation. In proposed multiplier as there is no need of register unit the time taken
to complete the execution is less and the number of clock cycles also gets
reduced. The proposed Booth multiplier occupies less area when compared
to "CSD" and "Booth multiplier." The latency of the proposed multiplier
is less compared to CSD as there is elimination of 4 delay elements in the
architecture due to which speed of the proposed multiplier is increased.

TABLE 3.7   Utilization of Hardware Components for 2D DWT Architecture

Multiplier	Column Filter				Transpose Unit	Row Filter			
	XOR/ XNOR	AND/OR/ NOR/NAND	Register	Latency	Register	XOR/ XNOR	AND/OR/ NOR/NAND	Register	Latency
CSD	768	1920	35	12	3	768	1920	35	12
BOOTH	1164	2536	17	8	3	1164	2536	16	8
Proposed	802	1949	17	8	3	802	1949	16	8

3.5   SYNTHESIS RESULTS

The synthesis results of "2D-DWT" architecture are done using "Verilog
HDL" in "Cadence Genus 90 nm technology," and the comparison table for
different methods with parameters like area, power, and delay are shown
in Table 3.8 for the "2D-DWT architecture." The observation in Table 3.8
shows three different methods like CSD, existing booth multiplier, proposed
Booth multiplier, which shows improvement in "area-delay-product (ADP)"
and "power-delay-product (PDP)."

TABLE 3.8   Synthesis Results for 2D DWT

Method	Delay (ns)	Area (μm^2)	Power (mW)	ADP	EADP (%)	PDP	EPDP (%)
CSD	1.522	42219.1	1.78	64257.5	29.0194	2.713187	26.12955
BOOTH	1.862	34297.4	1.56	63861.8	28.2248	2.903482	34.97588
Proposed	1.580	31521.9	1.36	49804.5	–	2.151112	–

The "excess-area-delay-product (EADP)" and "excess-power-delay-product (EPDP)" give the relation how much excess is that method by other method. The "ASIC implementation" results show the ADP of CSD representation is 29.01% in excess and for existing "Booth multiplier" it is 28.22% in excess than proposed multiplier. The "PDP" of "CSD" is 26.13% excess and 34.98% in excess than proposed multiplier. The above results clearly show that the proposed multiplier has less area and low power than the CSD and existing "Booth multiplier" because the reduction in generation of partial products is made; only the necessary product terms are generated. The above synthesis results make a clear detail applying proposed multiplier for picture handling capabilities and fast speedy applications.

The delay is not improved in the proposed method because in CSD the "critical path" is only due to adders in the circuit and it is from input to output. In the proposed method, the critical path is taken from the truncated part to the output which increases the delay in the circuit. Figures 3.14 and 3.15 shown tells that the proposed method has less area and power than the CSD and Booth multiplier due to reduction in minimization of partial products generation, only the required terms are generated so that area gets reduced. A bar graph representation of different multiplier methods is represented along X-axis and area in terms of sq-m is shown in Figure 3.14 in which shows proposed multiplier has low area. Similarly, in Figure 3.15 shows the bar-graph representation for different methods along X-axis and power in terms of mW along Y axis. These results are helpful for fast speed implementations and capable for picture handling applications.

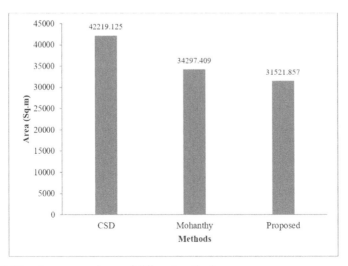

FIGURE 3.14   Area comparisons of different methods.

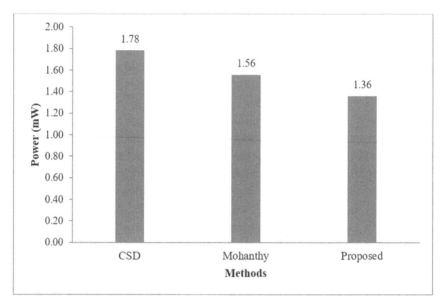

FIGURE 3.15   Power comparisons of different methods.

3.6   CONCLUSION AND FUTURE SCOPE

A "16-bit Radix-8 Booth multiplier" is designed and applied it to the "2D-DWT" architecture with the aim to have low area and power. With comparison to "CSD" and "Booth multiplier" the proposed multiplier has low area and power due to reduction in a number of partial product terms. In a "2D-DWT" architecture when enhancements are made by eliminating the additional defer components their watches a smaller number of clock term cycles than the "2D-DWT" with "CSD" portrayal and for a nitty gritty yield with no interruption of the worth is finished. On performing union, the proposed technique devours less territory and force, accordingly sparing of 29.01% "EADP" and 26.12% "EPDP" for existing "CSD" based engineering and 22.25% "EADP" and 34.97% "EPDP" for existing corner multiplier is accomplished with a mistake pace of +/−1%. This one level "2D-DWT" engineering can be stretched out to stagger "2D-DWT" as the future work to improve the plan.

KEYWORDS

- **convolution**
- **discrete wavelet transform**
- **fast Fourier transform**
- **filter**
- **flipping**
- **lifting**

REFERENCES

1. Huang, C. T., Tseng, P. C., & Chen, L. G., (2004). Flipping structure: An efficient VLSI architecture for lifting-based discrete wavelet transform. *IEEE Trans. on Signal Processing, 52*(4), 1080–1089.
2. Lai, Y. K., Chen, L. F., & Shih, Y. C., (2009). A high-performance and memory-efficient VLSI architecture with parallel scanning method for 2-D lifting-based discrete wavelet transform. *IEEE Trans. Consum. Electron., 55*(2), 400–407.
3. Darji, R. A., Merchant, S. N., & Chandorkar, A., (2014). Multiplier-less pipeline architecture for lifting-based two-dimensional discrete wavelet transform. *IET Computers & Digital Techniques, 9*(2), 113–123.
4. Meher, P. K., Mohanty, B. K., & Swamy, M. N. S., (2015). Low-area and low-power reconfigurable architecture for convolution-based 1-D DWT using 9/7 and 5/3 filters. In: *Proc. of Int. Conf. VLSI Design* (pp. 327–332).
5. Mohanty, B. K., & Meher, P. K., (2011). Memory-efficient modular VLSI architecture for high-throughput and low-latency implementation of multilevel lifting 2-D DWT. *IEEE Trans. Signal Process., 59*(5), 2072–2084.
6. Hu, Y., & Jong, C. C., (2013). A memory-efficient scalable architecture for lifting-based discrete wavelet transform. *IEEE Trans. Circuit Syst. II Express Brief, 60*(8), 502–506.
7. Mohanty, B. K., & Meher, P. K., (2013). Memory efficient high speed convolution based generic structure for multilevel 2D DWT. *IEEE Trans. Circuits Syst. Video Technol., 23*(2), 353–363.
8. Mohanty, B. K., & Meher, P. K., (2014). Area-delay-power-efficient architecture for folded two-dimensional discrete wavelet transform by multiple lifting computation. *IET Image Process, 8*(6), 345–353.
9. Acharya, T., & Chakrabarti, C., (2006). A survey on lifting-based discrete wavelet transform architectures. *Journal of VLSI Signal Processing Systems for Signal, Image and Video Technology, 42*(3), 321–339. doi: 10.1007/s11266-006-4191-3.
10. Kuang, S., Wang, J., & Guo, C., (2009). Modified Booth multiplier with a regular partial product array. *IEEE Trans. Circuits Syst. II, 56*(5), 404–408.
11. Jiang, H., Han, J., Qiao, F., & Lombardi, F., (2016). Approximate radix-8 Booth multipliers for low-power and high-performance operation. *IEEE Trans. Comput., 65*(8), 2638–2644.

12. Wu, C., Zhang, W., & Liu, J., (2017). Hardware efficient multiplier-less multi-level 2D DWT architecture without off-chip RAM. *IET Image Processing, 11*, 362–369.

13. Hewlitt, R. M., & Swartzlantler, E. S., (2000). Canonical signed digit representation for FIR digital filters. *2000 IEEE Workshop on Signal Processing Systems; SiPS 2000*. Design and Implementation (Cat. No.00TH8528). doi:10.1109/sips.2000.886740.

14. Mohanty, B. K., Mahajan, A., & Meher, P. K., (2012). Area and power-efficient architecture for high-throughput implementation of lifting 2-D DWT. *IEEE Trans. Circuits Syst. II Express Briefs, 59*(7), 434–438.

15. Darji, S. A., & Oza, A., (2014). Dual-scan parallel flipping architecture for a lifting-based 2-D discrete wavelet transform. *IEEE Trans. Circuit Syst. II Express Brief, 61*(6), 433–437.

16. Zhang, W., Jiang, Z., Gao, Z., & Liu, Y., (2012). An efficient VLSI architecture for lifting-based discrete wavelet transform. *IEEE Trans. Circuits Syst. II Express Briefs, 59*(3), 158–162.

17. Mohanty, B. K., & Choubey, A., (2017). Efficient design for radix-8 booth multiplier and its application in lifting 2-D DWT. *Circuits Syst. Signal Process, 36*(3), 1129–1149.

CHAPTER 4

Design and Performance Evaluation of Energy Efficient 8-Bit ALU at Ultra-Low Supply Voltages Using FinFET with 20 nm Technology

VALLABHUNI VIJAY,[1] PITTALA CHANDRA SHEKAR,[2] SHAIK SADULLA,[3] PUTTA MANOJA,[1] RALLABHANDY ABHINAYA,[1] MERUGU RACHANA,[1] and NAKKA NIKHIL[1]

[1]*Department of Electronics and Communication Engineering, Institute of Aeronautical Engineering, Dundigal – 500043, Hyderabad, Telangana, India, E-mail: v.vijay@iare.ac.in (V. Vijay)*

[2]*Department of Electronics and Communication Engineering, MLR Institute of Technology, Dundigal – 500043, Hyderabad, Telangana, India*

[3]*Department of Electronics and Communication Engineering, KKR, and KSR Institute of Technology and Sciences, Vinjanampadu, Andhra Pradesh – 522017, India*

ABSTRACT

In the last few years, the tiny size of MOSFET (i.e., less than tens of nano-meters) created some operational problems such as increased gate-oxide leakage, amplified junction leakage, high sub-threshold conduction, and reduced output resistance. To overcome the above challenges, FinFET has the advantages of an increase in the operating speed, reduced power consumption, decreased static leakage current is used to realize the majority of the applications by replacing MOSFET. By considering the attractive features of the FinFET, an ALU is designed as an application. In the digital processor, the arithmetic and logical operations are executed using the arithmetic logic unit (ALU). In this chapter, power-efficient 8-bit ALU is designed with full adder (FA) and multiplexers composed of gate diffusion input (GDI) which gained the designer's choice for digital combinational circuit realization at minimum power consumption. The design is simulated using Cadence virtuoso with 20 nm technology. Comparative performance analysis is carried out in contrast to the other standard circuits by taking the critical performance metrics such as delay, power, and power delay product (PDP), energy-delay product (EDP) metrics into consideration.

4.1 INTRODUCTION

The limits of the CMOS technology advancements, specifically, reducing the channel length of the MOS transistor, causing the inefficiency in realizing the optimized low power circuits at room temperature because of the 60 mV/ decade threshold swing limit. The published research content over the past few years in the literature convincing the FinFET as the most reliable device for the digital applications. The development and realization of CMOS technology-based circuits at ultralow supply voltages for the next level ICs is the current challenge for the design engineers. This chapter offers the circuit's sensitivities and the collaborations with the evolving technology of FinFET, for realizing fundamental module of ALU. Primarily, at circuit level, FinFET based 1-bit and 8-bit ALUs blocks are realized and tested by considering the attractive features of FinFETs. The specter models of Si FinFET are considered at 20 nm technology for benchmarking and evaluating the performance metrics of proposed designs. The designs are capable of offering optimized power levels by maintaining the reliability at utmost levels in contrast to that of the standard designs based on CMOS. The FinFET based designs are attaining the great choice by showing the consistent metrics like increased logic swing, diminished glitches, condensed overshoots, etc.

The very-large-scale integration (VLSI) is the task of designing an integrated circuit (IC) by clubbing thousands of transistors into one chip. The VLSI turned up in the year 1970 when the complex semiconductor and communication technologies were being emerged. The electronics industry achieved a considerable expansion over the last few decades, mainly because of the rapid promotions in the large-scale integration technologies and the applications of system design. With the advantages of the VLSI designs, the implementation of ICs in communications, video, and image processing has been rising. Even for no portable devices, power consumption is an essential criterion due to raising in the packaging and cooling costs along with reliability issues. Considering these problems, the primary challenge for the design engineers is inadequate, realization within a targeted power without compromising the performance requirement [1].

The tiny size of the MOSFET, less than tens of nanometers, created some operational problems such as high sub-threshold conduction, which means; in the MOSFET, the applied voltage to gate terminal is to be decreased to maintain the reliability [2–4]. Then the threshold voltage (Vth) which is to be applied to the MOSFET must also be reduced. As this Vth is decreased possibly to a very great extent, the transistor will not be able to switch from the complete turn-off to complete turn-on state and vice versa. The rise of gate-oxide leakage refers to the gate oxide, serving as an insulator in between the channel and gate must be designed as thin as possible in order to elevate the conductivity of the channel and performance of the device when it is in ON state and to decrease the subthreshold leakage during OFF state of the device. The increased junction leakage means the design of junction becomes quite more difficult in smaller devices, leading to higher doping levels which lead to drain induced barrier lowering. And many more drawbacks such as lower output resistance, lower transconductance.

The scaling of gate lengths to very short distances is possible with FinFET technology. FinFET-DGCMOS and conventional CMOS fabrication are similar to one another, except negligible disturbances, allows the participation of more efforts in the significant involvement of the yielding the applications. FinFET channel is an un-doped structure of smaller in size portion which is placed in a perpendicular direction to that of substrate terminal. Coulomb scattering is eliminated by un-doped channel because of faster mobility in FinFETs by the unwanted impurities [5]. The two gates of the FinFET are provided to control the short channel effects without aggressively scaling down the gate-oxide thickness and increasing the channel doping density. In terms of the structure of FinFET, it is a double gate device, and the gates are formed at the vertical side of the fin using a thin gate oxide layer. Fin is a thin layer between the source and drain. FinFET reduces the levels of the leakage current and mitigates the short channel effects [6]. In this chapter, the reduction in power consumption by scaling down the size of a transistor in nanoelectronics with the help of FinFET is achieved.

This chapter mainly focuses on the implementation of 8-bit ALU using FinFET with 20 nm technology. ALU is the core part of the central processing unit (CPU) and is a combinational digital electronic circuit which accomplishes all the logical and arithmetic operations. Arithmetic modules provide minimum energy consumption at extended reliability plays a crucial role in the die. Thereby, the design changes in the ALU reflect a significant effect on the overall performance of the whole processor. To meet the reduced energy consumption and increased speed of ALU, designs require arithmetic, circuit, and system-level methodologies involvement [7]. Even much of the logic styles viz. clock gating, frequency, and voltage scaling are producing the minimum power consumption applications, but few amongst those are offering improved power consumption factor with the optimized die size [9].

The emerging features in the new era of IC are driven with the modern days' technological advancements in the processing of silicon technology. With the revolutionary change in the realization of commercial ICs, the chip development foundries are producing the highly complex, enhanced reliability, minimum energy consumption, and more robustness processors. The CPU is the basic module inside the microprocessor and composed with the ALU as a primary module. The key operations of the ALU are both arithmetic and logical operations. The arithmetic operations of addition, subtraction, multiplication, and division are performed as addition, inverted addition, recursive addition, and recursive inverted addition, respectively. The majority of the addition operations in the digital system are formed using the full adder, and the major constraint in the full adder design is to produce the circuits with minimum energy dissipation, less delay, energy efficiency in addition to reliability feature [10]. In contrast to the existed MOSFET designs, the novel innovation of FinFET models is considered to implement the full adder with the improved parameters of both device and energy efficiency. The 8-bit ALU is developed using Full adder and Multiplexers which are composed by using gate diffusion input (GDI) technique.

4.2   FINFET CHARACTERISTICS AND MODELING

FinFET has the name as such because of the field-effect transistor (FET) with a structure that view as a set of fins when gazed. It consists of a conducting region, which mainly surrounded with the thin 'fin' structure and is built on silicon on insulator by which the name 'FinFET' is evolved. The device channel for effective conduction is calculated with the fin thickness.

The FinFET is a non-planar device as shown in Figure 4.1; double gate transistor, which is either a bulk silicon-on-insulator (SOI) or on Silicon Wafers [2–11]. This is importantly based on single gate transistor design. There are two different types of FinFET. They are Bulk FinFET and SOI

FinFET. The critical characteristic of the FinFET is that it has a conducting channel that is bounded by a thin silicon fin [13]. This mainly forms the body of the device. These fins are nothing but the channel between the source and drain. The gate terminal is bounded around the channel. This lets the formation of the several gate electrodes to reduce the leakage current and to improve the drive current [14–16].

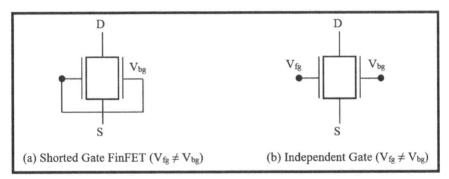

(a) Shorted Gate FinFET ($V_{fg} \neq V_{bg}$)   (b) Independent Gate ($V_{fg} \neq V_{bg}$)

FIGURE 4.1 FinFET models.

The FinFET works the same as that of Conventional MOSFET. It operates in two modes: (i) enhancement mode and (ii)depletion mode. The working characteristics are identical in both modes, but the only difference is that, in the enhancement mode, if no voltage is given to the gate terminal, it does not conduct whereas in the depletion mode, if the voltage is applied to the gate, it does not conduct. In the enhancement mode when the voltage is applied to the gate terminal, a parallel plate capacitor is formed. The gate is made up of the oxide layer. The surface below the oxide layer is located between the source and drain. When a small amount of positive voltage is applied to the gate concerning the source, a depletion region is formed. This region is reversed to n-type by the applied positive voltage. Then a region is formed at the interface between Si and SiO_2. This applied positive voltage attracts the electrons from the source terminal to the drain terminal. By this, an electron reach channel is formed. The flow of current starts by applying a voltage between the source and drain. This flow of current is dependent on the voltage applied to the gate.

The advantages and applications of FinFET are: it consumes very less power that allows higher integration levels [17, 18]; it operates at the lower voltages as the Vth given is less. The FinFET has passed the barrier of 20 nm which was previously a barrier. The static leakage current has been reduced up to 90% when compared to that of traditional methods, and its operational speed has been increased up to 30% than non FinFET devices. Samsung Electronics has incorporated FinFET in its 14 nm process. Along with Samsung, Apple, Intel, and Taiwan Semiconductor Manufacturing Company (TSMC) are set to ship the 14 nm technology, which is a benefit to all smartphones as it will speed with the phone.

4.3 PROPOSED 8 BIT ALU BLOCK USING FINFET MODELS

Apart from the traditional CMOS design, the GDI technique is another low power and area efficient technique [18, 19]. The aim of the GDI technique is to improve the performance of logic circuits [20, 21]. A common GDI cell comprises of four terminals. They are P (PFinFET outer diffusion node), G (Common gate input for both PFinFET and NFinFET transistors), D (Common diffusion for both the transistors), N (NFinFET outer diffusion node).

4.3.1 GDI-BASED MULTIPLEXER

GDI based logic is used to implement 2:1 MUX as shown in Figure 4.2, whose function is based on the use of a simple cell. When the select line is fed with '0', the output will be the data 'A.' And when the select line is fed with '1', the output will be the data 'B.' The truth Table 4.1 shows the functionality of the 2×1 Multiplexer.

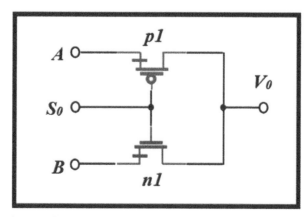

FIGURE 4.2   2×1 Multiplexer.

TABLE 4.1 The Truth Table of 2×1 Multiplexer

Select line	$V_0 = \overline{S_0}A + S_0 B$
S_0	Y
0	A
1	B

GDI based logic is to implement 4:1 MUX as shown in Figure 4.3, whose function is based on the use of a simple cell. When the select line S0 is fed with '0' the output will be the data 'A' or 'C.' And when select line S1 is fed with '1' the output will be the data 'B' or 'D.' Table 4.2 shows the functionality of 4×1 MUX.

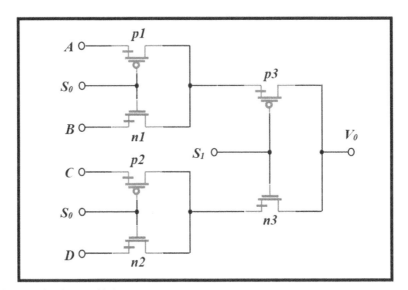

FIGURE 4.3 4×1 Multiplexer.

TABLE 4.2 The Truth Table of 4×1 Multiplexer

Select Line		Output
S_0	S_1	$V_0 = \overline{S_0}\,\overline{S_1}A + \overline{S_0}\,S_1 B + S0\,\overline{S_1}\,C + S_0 S_1 D$
0	0	A
0	1	C
1	0	B
1	1	D

4.3.2   11T FULL ADDER

A full adder is a fundamental requisite to design an ALU. The 11T methodology is used for developing Full Adder; this is another way of designing Full Adder to reduce power and delay. Full Adder is depicted in Figure 4.4. This circuit operates at very low voltages, such as less than 1V. This circuit is driven with a supply voltage of 0.9 (Vdd). The Input A is given to the gate of the terminal of p1, n1 and drain terminal of p2. Input B is applied to the gate terminals of the transistors p2 and n2 and the drain terminal of the n1. The passage of the signal occurs from gate to drain terminal during the ON condition of, and it became possible when the applied voltage (Vs) is higher than that of Vth. That means the voltage is passed from gate to drain. So, when the Input A is high, it gives the signal applied to the terminal B. Full Adder is built using low power EX-OR gates and 2×1 Multiplexer. The outputs of 'Sum' and 'carry' (Cout) are obtained from EX-OR and Multiplexer circuits, respectively. Additional transistor numbered as n6 takes minimum power as it is driven with the ultra-low mode in combination with the sub-threshold current. The voltage across the gate to source (VGS) become higher than that of Vth because of strong inversion region, and this leads to accumulating minority charge carriers by vanishing the majority charge carriers. The subthreshold current being generated because of leakage current by the deployment of minority charge carriers during the case of the weak inversion region of VTH is greater than VGS. The required operation of the circuit is met with sufficient current which is gained from the subthreshold current at VDD value is lower than the VTH. The resultant current gives the circuit to operate with ultra-low mode by consuming minimum power. Successful operation of 11T full adder is achieved with an additional transistor, n6 at the subthreshold mode. The whole operations of the circuit can be validated as shown in Table 4.3.

TABLE 4.3   The Truth Table of 11T Full Adder

Input			Output	
A	B	Cin	Sum	Cout
0	0	0	0	0
0	0	1	1	0
0	1	0	1	0
0	1	1	0	1
1	0	0	1	0
1	0	1	0	1
1	1	0	0	1
1	1	1	1	1

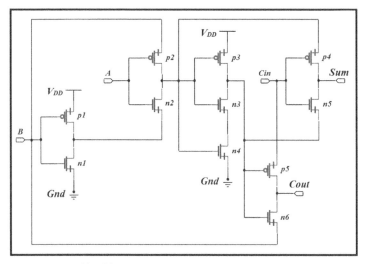

FIGURE 4.4   11T Full adder.

4.3.3   1-BIT AND 8-BIT ALU

For any input over the input terminals A, B, and C the ALU perform arithmetic and logical operations depending on the input given over the select line. The ALU is a part of the processor that tackles all the arithmetic and logical operations as it is named. The truth Table 4.4 for Figure 4.5, shows the operations performed by the ALU depend on 3 Bit select line giving eight conditions.

TABLE 4.4   1-Bit ALU for the Inputs A = 0, B = 1, and C = 1

Select Lines			Operation	Output
S_2	S_1	S_0		
0	0	0	AND	0
0	0	1	INCREMENT	1
0	1	0	OR	1
0	1	1	DECREMENT	0
1	0	0	SUBTRACTION	1
1	0	1	XOR	0
1	1	0	SUBTRACTION	0
1	1	1	ADDITION	1

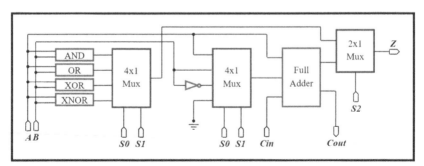

FIGURE 4.5   Block diagram of 1-bit ALU.

Let the input A is driven with logical '0,' B with logical '1' and C with logical '1.' The Logical AND operation is performed when the select line S_2 is fed with logical '0,' S_1 logical '0' and S_0 as logical '0.' The ALU is INCREMENTED from its present state when the select line S_2 is driven by logical '0,' S_1 with logical '0' and S_0 logical '1.' The operation Logical OR is implemented when S_2 is fed with logical '0,' S_1 logical '1' and S_0 as logical '0.' DECREMENT operation is carried out when select line S_2 is fed with logical '0,' S_1 logical '1' and S_0 with logical '1.' The SUBTRACTION operation turns up for the applied signal over select line S_2 is logical '1,' S_1 is logical '0,' and S_0 is logical '0.' For the Logical EX-OR operation the select line S_2 is to be fed with logical '1,' S_1 logical '0' and S_0 logical '1.' The Logical EX-NOR operation is accomplished when the select line S_2 is fed with logical '1,' S_1 with logical '1' and S_0 with a logical '0.' That which is last, the ADDITION operation is carried out for S_2 is fed with logical '1,' S_1 with logical '1' and S_0 with logical '1.'

For the 8-Bit ALU shown in Figure 4.6, A, B, and C are Inputs of ALU. S_0, S_1, and S_2 are 3-bit selection lines. F0 and F1 are Output of 1-bit ALU. F14 is the output of 8-Bit ALU, and F15 is carry obtained for last Full adder.

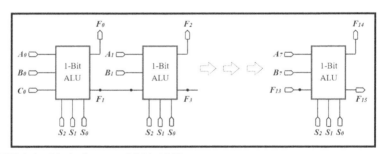

FIGURE 4.6   Block diagram of 8-bit ALU.

For any input over the input terminals A, B, and C, the ALU performs arithmetic and logical operations depending on the input given over the select line. The 8-bit ALU is obtained by cascading all 1-Bit ALU's [22–25]. The above block diagram shows the cascaded form of all 1-Bit ALU's. The carry which is obtained for the first Full adder's ALU is forwarded to the next Full adder's ALU. This process is continued to last ALU to form an 8-bit ALU.

4.4   SIMULATION RESULTS AND EXPERIMENTAL FINDINGS OF THE PROPOSED WORK

The circuit is shown in Figures 4.2 and 4.3 are simulated using the Cadence virtuoso tool by considering the FinFET technology specter files. The voltage applied to the proposed circuit is 600 mv or 0.6 V which falls under the operating voltage range of the FinFET. The multiplexers are designed using GDI technique which is advantageous in the field of reducing the circuit area and the supply voltage. The output (Vout) is obtained depending on the selection lines (S), (S$_0$) and (S$_1$) for both the multiplexers. When (S) is given with 0 V, the input 'A' is passed to the output, and when (S) is given with 0.6 V then 'B' is passed to the output. Similarly, when (S$_0$) is applied with 0 V the input 'A' and 'C' are passed to the next level, for (S$_1$) when applied with 0.6 V, input 'B' and 'D' is passed to the next level. Depending on data over terminal Vs1 the inputs 'A,' 'B,' 'C' or 'D' as passed over the output terminal. The simulated output of Figures 4.2 and 4.3 are shown in Figures 4.7 and 4.8, respectively.

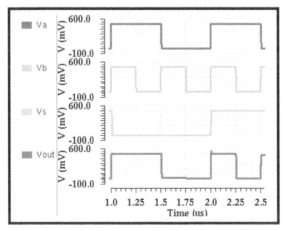

FIGURE 4.7   Simulated response of 2×1 MUX shown in Figure 4.2.

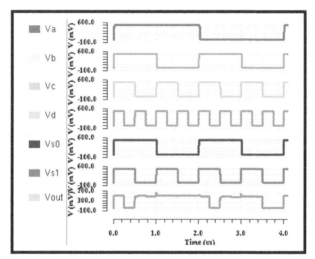

FIGURE 4.8   Simulated response of 4×1 MUX shown in Figure 4.3.

Full Adder using 11T is shown in Figure 4.4. It is simulated using the Cadence Virtuoso tool by considering the FinFET technology spectra files. Moreover, its simulated output is shown in Figure 4.9. The 11T Full Adder operates in ultra-low mode by consuming minimum power. The voltage applied to the proposed circuit is 600 mv or 0.6 V. The Full Adder is responsible for the three-bit addition, whereas Half Adder can only add two bits and neglects the carry which is obtained. The input (C) is added along with the bits with (A) and (B) which is the carry output obtained from the circuit itself.

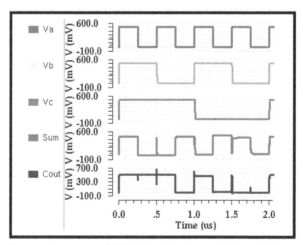

FIGURE 4.9   Simulated response of 11T full adder shown in Figure 4.4.

The simulation of 1-Bit ALU shown in Figure 4.5 is carried out and the simulated output is depicted in Figure 4.10. The proposed circuit operates with a voltage of 0.6 V or 600 mv. The response consists of eight different operations, which are the combination of arithmetic and logical operations and these operations depend on the input given over the selection lines (S_0), (S_1) and (S_2). The (F0) is the output of the 1-Bit ALU as shown in Figure 4.10 and (F1) is the input which is applied to the ALU during cascading.

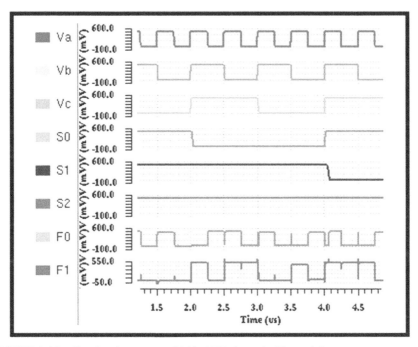

FIGURE 4.10 Simulated response of 1-bit ALU shown in Figure 4.5.

The above are the transient responses of 8 individual operations performed by the 1-Bit ALU. The ALU yields eight different operations in which four are arithmetic, and four are logical operations depending on the select line. Figure 4.11 shows the different operations of the 1-Bit ALU for the supply voltage of 0.6 V. The role of selection line which is about three bit plays a vital role in delivering the output responses which are as follows.

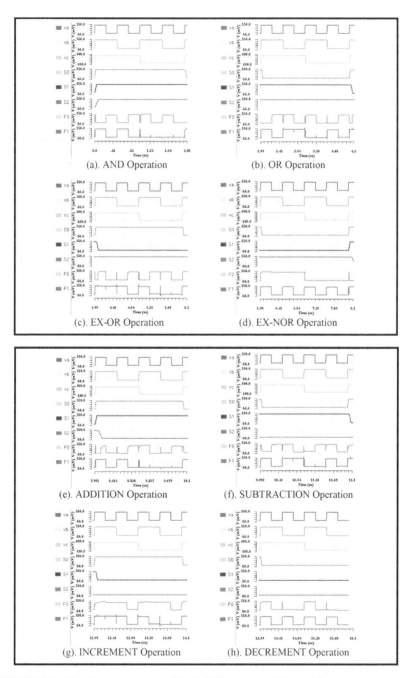

FIGURE 4.11   Simulated responses of 1-bit ALU shown in Figure 4.5.

Figure 4.11(a) shows logical AND operation when the select line S_2 is fed with logical '0,' S_1 logical '0' and S_0 as logical '0.' The ALU is INCRE-MENTED from its present state when the select line S2 is driven by logical '0,' S_1 with logical '0' and S_0 logical '1' as shown in Figure 4.11(g). The Logical OR Operation as shown in Figure 4.11(b) is implemented when S_2 is fed with logical '0,' S_1 logical '1' and S_0 as logical '0.' The DECREMENT operation is carried out when select line S_2 is fed with logical '0,' S_1 logical '1' and S_0 with logical '1' as depicted in Figure 4.11(h). As shown in Figure 4.11(f) SUBTRACTION operation turns up for the applied signal over select line S_2 is logical '1,' S_1 is logical '0,' S_0 is logical '0.' For the Logical EX-OR operation in Figure 4.11(d) the select line S_2 is to be fed with logical '1,' S_1 logical '0' and S_0 logical '1.' The Logical EX-NOR operation is accomplished when the select line S_2 is fed with logical '1,' S_1 with logical '1' and S_0 with a logical '0' as shown in Figure 4.11(c). Also, Figure 4.11(e) shows the ADDITION operation is carried out for S_2 is fed with logical '1,' S_1 with logical '1' and S_0 with logical '1.'

The simulated output of 8-Bit ALU as shown in Figure 4.12 is obtained by simulating the circuit shown in Figure 4.6. The voltage applied to the proposed circuit is 0.6 V. The same selection input is applied to all the 1-Bit cascaded ALU's along with the supply voltage. For the different inputs over the selection line, the proposed circuit performs eight different operations, including four arithmetic and four logical operations.

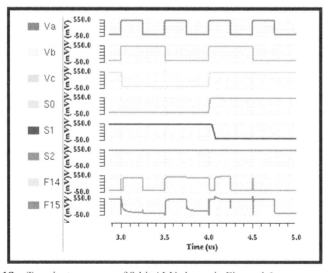

FIGURE 4.12   Transient response of 8-bit ALU shown in Figure 4.6.

Table 4.5 shows the obtained results by simulating the 8-bit ALU using FinFET in the cadence virtuoso tool. The 8-bit ALU is designed using Full adder and multiplexers. The key metrics of power, delay, PDP as well as EDP are analyzed for multiplexers, Full adder, 1-bit ALU and for 8-bit ALU.

TABLE 4.5   Analysis of Power, Delay, PDP, and EDP

Design	Power (nW)	Delay (ns)	PDP ($\times 10^{-18}$ J)	EDP ($\times 10^{-30}$ Js)
Full adder	3.6	0.0044	0.01584	0.069
GDI based 2×1 MUX	9.9	0.0013	0.01287	0.016
GDI based 4×1 MUX	19	8.1	153.9	1246590
1-Bit ALU	26.8	2.17	58.15	126198
8-Bit ALU	200	1.97	394	776180

The results show that the power consumed for the Full adder is 3.6 nW, GDI based 2×1 MUX, and GDI based 4×1 MUX is 9.9 nW and 19 nW, respectively. For the 1-bit ALU it is 26.8 nW and 8-bit ALU it is 0.2 μW. The delay metrics calculated for Full adder is 4.4ps, GDI based 2×1 MUX, and GDI based 4×1 MUX is 1.3ps and 8.1ns, respectively. For the 1-bit ALU and 8-bit ALU, the delay metrics are 2.17ns and 1.97ns. The product of power and delay is the standard performance parameter which is taken into consideration when comparing standard circuits. The PDP for Full adder is 15.84×10^{-21} J, GDI based 2×1 MUX is 12.87×10^{-21} J and GDI based 4×1 MUX is 153.9×10^{-18} J, 1-bit ALU and 8-bit ALU are 58.15×10^{-18} J and 394×10^{-18} J, respectively. The product of energy and delay is also another significant factor which is to be measured. The energy-delay product for full adder is 0.069×10^{-30} Js, GDI based 2×1 MUX is 0.016 Js, and GDI based 4×1 is 124.659×10^{-27} Js, 1-bit ALU and 8-bit ALU are 126.198×10^{-27} Js and 776×10^{-27} Js, respectively.

4.5   CONCLUSION

In this chapter, the 8-bit ALU is designed using GDI based 4×1 and 2×1 multiplexers and a Full adder. As CMOS has some operational problems like short channel effects which include Drain-induced barrier lowering, high sub-threshold conduction, increased gate oxide leakage, increased junction leakage, lower output resistance. To overcome the above problems,

CMOS is replaced with FinFET. ALU represents the fundamental building block of the CPU of a computer, and modern CPUs contains mighty and complex ALU. It performs arithmetic operations, which includes addition, subtraction, increment, decrement, and logical operations, which includes OR, AND, XOR, and XNOR. The required gates for these operations are designed using FinFET, and a multiplexer performs logical operations. Here, the power consumption, circuit's delay, in addition to their PDP along with EDP are analyzed which gave the quite better values in terms of power and delay when compared to CMOS technology. The delay of 8-bit ALU is minimized to 1.97ns from 6.95ns and the power consumption reduced to 0.2 µwatts from 32 µwatts when compared to that of CMOS technology. The FinFET based circuit gave the required output by reducing the delay, power consumption at ultra-low supply voltages.

KEYWORDS

- **8-bit ALU**
- **central processing unit**
- **field-effect transistor**
- **full adder**
- **gate diffusion input**
- **integrated circuit**
- **static leakage current**

REFERENCES

1. Shiva, T., & Rahebeh, N. A., (2013). Aging comparative analysis of high-performance FinFET and CMOS flip-flops. *Microelectronics Reliability, 69*(2), 52–59.
2. Osama, A., El-Din, M. M., Hassan, M., Hamdy, A., Hossam, A. H. F., Yehea, I., & Ahmed, M. S., (2018). Technology scaling roadmap for FinFET-based FPGA clusters under process variations. *Journal of Circuits, Systems and Computers, 27*(4), 1850056(1–32).
3. Jain, A., & Saxena, M., (2014). Optimization of leakage current and leakage power of FinFET based 6T SRAM cell. *International Journal of Engineering Technology, Management and Applied Sciences, 2*(3), 156–163.
4. Vallabhuni, V., & Avireni, S., (2017). A novel square wave generator using second-generation differential current conveyor. *Arabian Journal for Science and Engineering, 42*(12), 4983–4990. Springer.

5. Lin, X., Wang, Y., & Pedram, M., (2014). Stack sizing analysis and optimization for FinFET logic cells and circuits operating in the sub or near-threshold regime. In: *15th International Symposium on Quality Electronic Design* (pp. 341–348). Santa Clara CA, USA.

6. Hajare, R., Sunil, C., Kumar, A. R. A., Jain, P. R. S., & Sriram, A. S., (2015). Performance analysis of FinFET based inverter circuit, NAND and NOR gate at 22nm and 14nm node technologies. *International Journal on Recent and Innovation Trends in Computing and Communication, 3*(5), 2527–2532.

7. Nirmal, D., Vijaya, K. P., & Jabaraj, S., (2010). NAND gate using FinFET for nanoscale technology. *International Journal of Engineering Science and Technology, 2*(5), 1315–1358.

8. Vijay, V., & Avireni, S., (2016). A DCCII based square wave generator with grounded capacitor. In: *Proceedings of the 2016 IEEE International Conference on Circuits, Power and Computing Technologies (IEEE ICCPCT-2016)* (pp.1–4). Kumaracoil, India.

9. Aminul, I., Akaram, M. W., & Hasan, M., (2011). Energy-efficient and process tolerant full adder in technologies beyond CMOS. *Int. J. on Recent Trends in Engineering & Technology, 5*(1), 60–68.

10. Alak, M., & Pritam, B., (2018). Variation aware intuitive clock gating to mitigate on-chip power supply noise. *International Journal of Electronics, 105*, 1487–1500.

11. Wei, L., Huei, C. C., Cheng, S. L., & Michael, L. P. T., (2014). Performance evaluation of 14nm FinFET-based 6T SRAM cell functionality for DC and transient circuit analysis. *Journal of Nanomaterials, 2014*, 1–8.

12. Vijay, V., & Avireni, S., (2016). Grounded resistor and capacitor based square wave generator using CMOS DCCII. In: *Proceedings of the 2016 IEEE International Conference on Inventive Computation Technologies (IEEE ICICT-2016)*, (pp. 79–82). Coimbatore, India.

13. Moulai, K. A. N., Bouazza, A., & Bouazza, B., (2013). Corner effects sensitivity to fin geometry variations in tri-gate SOI-FinFET. *Journal of Electronic Devices, 18*, 1549–1552.

14. Onabjo, M., & Martinez, J. S., (2012). Process variation challenges and solutions approaches. *Analog Circuit Design for Process Variation-Resilient Systems-on-a-Chip* (pp. 9–30) Springer, Boston, MA.

15. Vijay, V., & Avireni, S., (2013). A square wave generator using single CMOS DCCII. In: *Proceedings of the 2013 IEEE International SoC Design Conference (IEEE ISoCC-2013)* (pp. 322–325). Busan, South Korea.

16. Liu, J. C., Mukhopadhyay, S., Wang, Y. F., Tsai, Y. S., Chen, S. C., Lee, J. H., Lu, R., et al., (2017). A novel bit-level characterization methodology to benchmark the FinFET based SRAM performance under the influence of leakage current. In: *2017 IEEE International Electron Devices Meeting (IEDM)* (pp. 12.5.1–12.5.4). San Francisco, CA, USA.

17. Vinay, K., Ravindra, K. S., & Madhav, M. P., (2018). A temperature compensated read assist for low vmin and high performance high density 6T SRAM in FinFET technology. In: *2018 31st International Conference on VLSI Design and 2018 17th International Conference on Embedded Systems (VLSID)* (pp. 447–448). Pune, India.

18. Vijay, V., & Avireni, S., (2015). Tunable resistor and grounded capacitor based square wave generator using CMOS DCCII. *International J. Control Theory and Applications, 8*, 1–11.

19. Akshay, D., Abhilash, K., & Sandeep, K. D., (2014). Design and implementation of area optimized ALU Using GDI technique. *International Journal of Innovative Research in Electrical, Electronics, Instrumentation and Control Engineering, 2*(3), 1169–1172.
20. Anagh, D., & Vigneswaran, T., (2018). Design of arithmetic and logic unit (ALU) using subthreshold adiabatic logic for low-power application. *VLSI Design: Circuits, Systems and Applications* (pp. 201–209). Springer, Singapore.
21. Vallabhuni, V., & Avireni, S., (2019). A low power waveform generator using DCCII with grounded capacitor. *Int. J. Public Sector Performance Management, 5*, 134–145.
22. Rani, T. E., Rani, M. A., & Rao, R., (2011). AREA optimized low power arithmetic and logic unit. In: *2011 3rd IEEE International Conference on Electronics Computer Technology (ICECT)* (pp. 224–228). Kanyakumari, India.
23. Polani, R., Korada, P. R., & Umadevi, S., (2018). 8-bit asynchronous wave-pipelined arithmetic logic unit. *Nanoelectronic Materials and Devices* (pp. 233–243) Springer, Singapore.
24. Vallabhuni, V., Siva, N. V., Sai, G. M., Revanth, R. B., Suresh, K. U., & Surekha, C., (2019). A simple and enhanced low-light image enhancement process using effective illumination mapping approach. *Lecture Notes in Computational Vision and Biomechanics* (pp. 975–984) Cham, Switzerland.
25. Ghosh, P., Saha, T., & Kumari, B., (2018). Aspects of low-power high-speed CMOS VLSI design: A review. *Industry Interactive Innovations in Science, Engineering and Technology* (pp. 385–394). Springer, Singapore.

CHAPTER 5

Design and Statistical Analysis of Strong Arbiter PUFs for Device Authentication and Identification

KURRA ANIL KUMAR and USHA RANI NELAKUDITI

Department of Electronics and Communications Engineering, Vignan's Foundation for Science, Technology, and Research (Deemed to be University), Guntur, Andhra Pradesh, India, E-mail: kakumar94@gmail.com (K. A. Kumar)

ABSTRACT

With the advancement of VLSI technology, people strongly rely on electronic devices. Integrated circuits (ICs) are becoming an integral part of any electronic device. Over the past decade, the semiconductor supply chain facing the several vulnerabilities like counterfeiting, cloning, tampering, etc. Hence to prevent semiconductor counterfeits from adversaries, a physical unclonable functions (PUFs) are emerged to be a one of the light weight security primitives in the hardware cryptography and offers the robust security against attacks. PUFs provide signatures it can act like a fingerprint for the each device. Due to its ability to generate the die specific unique identifiers, this can be used in chip authentication and identification protocol. Due to limited circuit resources, therefore generating a large set of Challenge-Response Pairs (CRP) data base results increases the cost. Hence to address this limitation proposed a multiplexer-decoder based arbiter PUF using CMOS 45 nm technology and simulated using cadence virtuoso. From the obtained results estimated security metrics like uniqueness, reliability, and randomness. Therefore, it is clear evident that proposed arbiter PUF is well suited for the detection and identification of cloned or counterfeited electronic devices.

5.1   INTRODUCTION

As day-by-day technology has been enormously increases then security places a big challenge in the current day electronic devices. The low-cost electronic devices face the several tough challenges like lack of computational resources, high power dissipation, cyber-attacks, counterfeiting, etc. To address these issues and to combat the above-mentioned problems PUFs are emerged to be one of the promising lightweight cryptographic primitives in the hardware security [1–4]. A PUF utilizes the unavoidable and uncontrollable manufacturing process variations and provides a unique information; it can be obtained by applying the random set of challenges and produces responses. Due to this process variations even same challenge being applied to different PUFs, it produces the unique responses it can be observed by Figure 5.2. The tamper sensitive property of a PUF can eliminate the vulnerabilities in secure storage and also avoids the storage of keys in expensive non-volatile memories. Counterfeiting the same PUF with the existing database is infeasible and practically it requires huge time and money [5–9]. Figure 5.1 depicts the generation of CRPs due to manufacturing process variations.

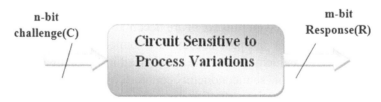

FIGURE 5.1   Generation mechanism of CRPs due to manufacturing variations.

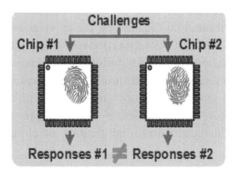

FIGURE 5.2   CRPs relationship with ICs.

A systematic spatial process variation during fabrication of integrated circuits (ICs) can alter the electrical behavior of the device and these can observe from die-to-die, wafer-to-wafer, lot-to-lot, and IC-to-IC, respectively. The typical complementary metal-oxide semiconductor (CMOS) process variations can be shown in Figure 5.3. During the fabrication of any IC, we can encounter two types of mismatches such as global mismatches and local mismatches, and it can be mathematically expressed by Eqn. (5.1).

$$M_{Total} = M_{Global} + M_{Local} \tag{5.1}$$

Global mismatches are inherently occurred and parameters like channel length, impurity concentration, oxide thickness, threshold voltage, diffusion depth, etc., are mainly responsible for these variations. On the other hand, parameters like supply voltage, aging, and environmental conditions are responsible for the occurrence of local mismatches, respectively [10–14]. The effect of the variations cannot be controlled completely and minimized. Global parameters are likely responsible for the generating different CRPs and local parameters effects the PUF circuit and reduces the yield in terms of producing the less uniqueness, reliability, and randomness. Especially aging can much impact and degrades the system performance mainly due to phenomenon like negative bias temperature instability (NBTI), positive bias temperature instability (PBTI), electromigration, hot carrier injection (HCI), time-dependent dielectric breakdown (TDDB), etc. [15–18]. According to the size challenge-response pair (CRP) space PUFs can be broadly classified in to weak and strong PUFs [19–23]. A weak PUF having limited number of CRPs, or its number of challenges increases linearly or polynomial with the number of fundamental blocks adjoined to form a PUF, weak PUFs normally used for random number and device key generation. In contrast, the number of CRPs of a strong PUF increases exponentially with the number of basic cells [24–28].

5.2 BACKGROUND

Over the several years, different PUF architectures being proposed and analyzed. It includes arbiter PUFs [29, 30], feed-forward PUFs [31], mux-demux PUFs [32], ring oscillator PUFs (RO-PUF) [33], glitch PUF [34], memory-based PUF [35], latch PUF and flip-flop PUF [36], etc. A quantity of work related to PUF architectures have been found in Refs. [37–40]. This section provides a basic overview about strong PUF architectures and its operation.

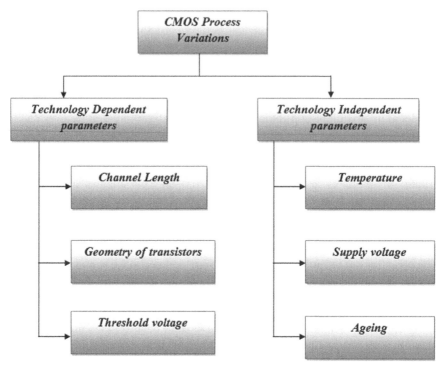

FIGURE 5.3 Classification of CMOS process variations.

5.2.1 OPERATION OF RING OSCILLATOR PUF

Suh and Devadas proposed a basic ring oscillator PUF [41] in 2007, due to its less complexity and ease of implementation it could be widely used as a true random number generator (TRNG). The basic architecture of the ring oscillator PUF as shown in Figure 5.4. A typical ring oscillator PUF having K-ring oscillators and 2k to 1 multiplexer. In this design, a set of ring oscillators are connected to the counter, and the final response of the counter is given as input to the comparator. The functionality of comparator is to count the oscillations generated by the ring oscillator over the period of time. Due to random process variations at each and every stage there will be change in the frequencies. Therefore, this can be figured out by using the counter and produces the response as a 0 or 1, respectively. An n bit signature can be generated by comparing the several ROs, and these can be implemented by using the macros and controlled by using the multiplexers. If $f_i < f_j$ (f_i and f_j are the frequencies of RO_i and RO_j oscillators, respectively) comparator output will be '0' otherwise '1.' The conventional RO PUF occupies the huge area and dissipates the power. To address the above-mentioned problems in ring oscillator PUFs, proposed another set of PUFs such as arbiter PUFs.

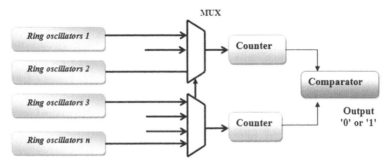

FIGURE 5.4   Ring oscillator PUF.

5.2.2   OPERATION OF ARBITER PUF

Figure 5.5 illustrates the basic structure of the arbiter PUF. This was first introduced by Gassend et al 2004 [42, 43]. This is also another type of strong PUF; it consists of a delay network and arbiter stage. The delay network consists of 'N' identical switching elements (multiplexers) which are connected at top and bottom stage. An arbiter can act as decision element which is fed at the final stage of the delay network and arbiter could be either SR latch or D latch, its converters the analog timing delay difference in to digital value. The output of the TRNG can be acts as challenges for the delay network, and it can be represented as challenge vector (C1, C2, C3,... Cn). Rising edge of pulse signal could be given as enabling of the arbiter PUF circuit and based on challenges the signal can be propagated either cross or straight and final timing delay difference signal can be fed to the arbiter, it decides the response could be either '1' or '0.'

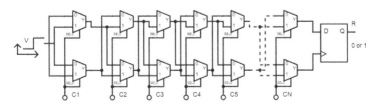

FIGURE 5.5   Schematic of the linear N-stage arbiter PUF.

The conventional arbiter PUF linearity in timing delay, hence it is very susceptible for the modeling attacks, to overcome this problem Gassend proposed feed forward arbiter PUF (as shown in Figure 5.6) adding arbiters at the intermediate stage of the conventional arbiter PUF, therefore the output of the arbiters can acts as the challenges for the following stages of the delay network, ultimately this configuration avoids the modeling attacks, however this could be degrades the reliability of the PUF circuit and also easily effected by environmental variation. To enhance the PUF metrics and avoid the attacks against the adversary proposed various types of arbiters PUFs like mux based reconfigurable PUFs, modified feed-forward arbiter PUFs, XOR arbiter PUF, etc. The architecture of the feed-forward and XOR arbiter PUF as shown in Figures 5.6 and 5.7, respectively.

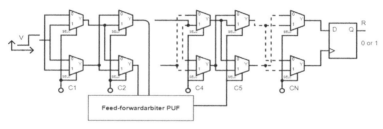

FIGURE 5.6   Feed-forward arbiter PUF architecture.

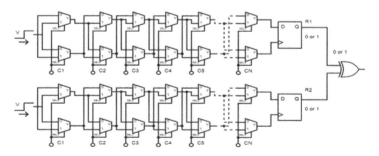

FIGURE 5.7   XOR arbiter PUF architecture.

5.3   ARCHITECTURE AND SIMULATION

5.3.1   *PROPOSED MUX-DECODER BASED 16-STAGE ARBITER PUF*

Figure 5.8 depicts the basic block diagram of the decoder mux-based arbiter PUF. The delay network composed of a decoder and multiplexer at each stage and SR latch can be used as arbiter. Due to two switching elements based on applied challenges the signal can travel in a random direction. The proposed 16-stage arbiter PUF being implemented in CMOS 45 nm technology using cadence specter. In order to assess the amount mismatches occurred in each stage performed the Monte Carlo (MC) analysis for 200 iterations and 150 samples. From the obtained results estimated security metrics (uniqueness, reliability, and randomness) and calculated its yield with respect to mean and standard deviation. Figure 5.9 depicts the schematics of the 16-stage decoder-mux based arbiter PUF.

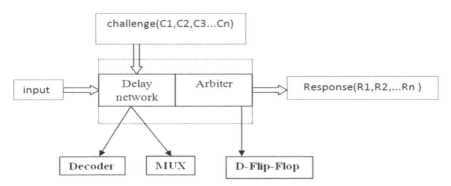

FIGURE 5.8   Block diagram of decoder-Mux based PUF structure.

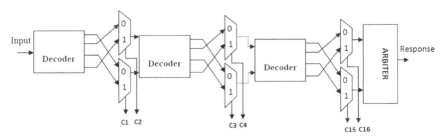

FIGURE 5.9   Schematic of the 16-stage decoder-mux based Arbiter PUF.

5.3.2   PUF SECURITY METRICS

5.3.2.1   UNIQUENESS

It is a measure of how differently the CRPs can be produced by a PUF from the other chips. This could be measured with the help of inter chip Hamming

Distance (HDinter). The ideal value of the uniqueness should be 50%. Mathematically it can be expressed by Eqn. (5.2):

$$\text{Uniqueness} = \frac{2}{k(k-1)} \sum_{k}^{k-1} \sum_{j=i+1}^{k} \frac{\text{HD}(R_i, R_j)}{N} \times 100\%$$

(5.2)

where; 'N' is the total number of PUF instances; R_i, R_j is two dissimilar chips i and j.

To estimate the uniqueness of the PUF, we applied the 200 iterations for 50 challenges and measured its mean and standard deviation from 3 sigma plot at different corners like slow-slow (SS), slow-fast (SF), fast-slow (FS), fast-fast (FF) and typical-typical (TT). All the responses are measured at room temperature (27°) at a 1v, respectively. From that obtained results the maximum uniqueness 48.22% from TT corner. Figure 5.10 depicts the percentage of uniqueness over different corners.

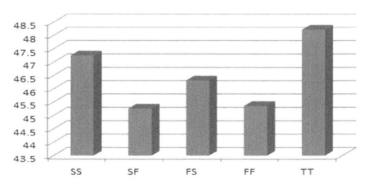

FIGURE 5.10   Variation of uniqueness over different corners.

5.3.2.2   RELIABILITY

It is the ability of PUF, how accurately the PUF can recreate the responses under various environmental conditions like voltage, temperature, and aging, etc. The ideal value of reliability is 100%, but practically its value is lies in between 95% and 99%, respectively, and it can be measured by using the Intra chip hamming distance (HDintra). Mathematically, it can be expressed by Eqn. (5.3):

$$\text{Reliability} = \frac{1}{s} \sum_{t=1}^{s} \frac{\text{HD}(R_i, R_{i,t})}{N} \times 100$$

(5.3)

To estimate the reliability of the PUF, chosen the reference temperature as 27°C and voltage as 1v. Applied a set of 50 random challenges and temperature ranges from –20 to 55°C. This process was carried out over 200 iterations over 150 samples. From the results we observed the reliability 91.28% and 92.58% with respect to supply voltage and temperature. Figures 5.11(a) and (b) represents the reliability of proposed PUF at different nodes of voltage and temperature at different instances.

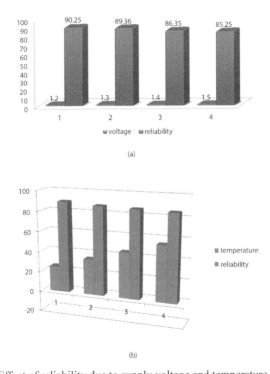

(a)

(b)

FIGURE 5.11 Effect of reliability due to supply voltage and temperature.

5.3.2.3 UNIFORMITY

It is a measure of randomness of a PUF, even with the prior idea of the PUF circuit the response generated from the PUF should be unpredictable for each CRP. The ideal value of uniformity could be 50% and mathematically it could be expressed by Eqn. (5.4).

$$\text{Uniformity} = \frac{1}{n}\sum_{i=1}^{n} R_{i,l} \times 100\% \qquad (5.4)$$

Figure 5.12 represents the distribution of the response of the PUF response over 150 number of samples. The test results indicate that the maximum mean could be 47.36%, respectively.

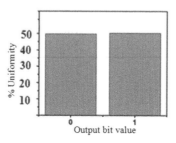

FIGURE 5.12   Estimation of uniformity ('0' and '1').

5.4   CONCLUSION

In this work, we demonstrated the different classifications of strong PUFs. In order to enhance the uniqueness, instead of using the convention multi-plexers, constructed PUF design using the multiplexer and decoder. As a result, the process variations can lead to a larger scale which makes the PUF can generate the unique signatures at each challenge. The proposed 16 stage mux-decoder based arbiter PUF are designed using the CMOS 45 nm technology and results are closely reaches the ideal value of the PUF metrics. And also, this architecture was tested over a wide range of voltages and temperatures. Hence this architecture well suitable for embedding PUF architectures in electronic devices to identify and authenticate the cloned or counterfeited devices.

KEYWORDS

- **hardware cryptography**
- **hot carrier injection**
- **Monte Carlo analysis**
- **negative bias temperature instability**
- **physical unclonable functions**
- **statistical analysis**

REFERENCES

1. Suh, G. E., & Devadas, S., (2007). Physical unclonable functions for device authentication and secret key generation. In: *Proc. 44th ACM Annual Design Automation Conf.* (pp. 9–14) San Diego, CA, USA.

2. Lee, J. W., Lim, D., Gassend, B., Suh, G. E., Van, D. M., & Devadas, S., (2004). A technique to build a secret key in integrated circuits for identification and authentication applications. In: *Proc. 2004 IEEE Symp. VLSI Circuits* (pp. 176–179). Honolulu, HI, USA.

3. Kumar, R., Patil, V. C., & Kundu, S., (2011). Design of unique and reliable physically unclonable functions based on current starved inverter chain. In: *Proc. 2011 IEEE Int. Symp. VLSI (ISVLSI)* (pp. 224–229). Chennai, India.

4. Liu, C. Q., Cao, Y., & Chang, C. H., (2017). ACRO-PUF: A low-power, reliable and aging-resilient current starved inverter-based ring oscillator physical unclonable function. *IEEE Trans. Cir. and Syst. I: Reg. Papers, 64*(12), 3138–3149.

5. Lin, L., Holcomb, D., Krishnappa, D. K., Shabadi, P., & Burleson, W., (2010). Low-power sub-threshold design of secure physical unclonable functions. In: *Proc. 16th ACM/IEEE Int. Symp. Low Power Electronics Design* (pp. 43–48). Austin, TX, USA.

6. Maes, R., Tuyls, P., & Verbauwhede, I., (2009). A soft decision helper data algorithm for SRAMPUFs. In: *Proceedings of the IEEE International Symposium on Information Theory* (pp. 2101–2105).

7. Hofer, M., & Boehm, C., (2010). An alternative to error correction for SRAM-like PUFs. In: *Proceedings of the International Workshop on Cryptographic Hardware and Embedded Systems* (pp. 335–350).

8. Chellappa, S., Dey, A., & Clark, L. T., (2012). SRAM-based unique chip identifier techniques. *IEEE Trans. VLSI Syst., 24*(4), 1213–1222.

9. Claes, M., Van, D. L. V., & Breaken, A., (2011). Comparison of SRAM and FF PUF in 65nm technology In: *Proceedings of the Nordic Conference on Secure IT Systems* (pp. 47–64).

10. Jang, J., & Ghosh, S., (2015). Design and analysis of novel SRAM PUFs with embedded latch for robustness. In: *Proceedings of the ISQED* (pp. 298–303).

11. Mathew, S. K., Satpathy, S. K., Anders, M. A., Kaul, H., Hsu, S. K., Agarwal, A., Chen, G. K., et al., (2014). A 0.19 pJ/b PVT-variation-tolerant hybrid physically unclonable function circuit for 100% stable secure key generation in 22 nm CMOS. In: *Proceedings of the ISSCC* (pp. 278–279).

12. Rosenblatt, S., Chellappa, S., Cestero, A., Robson, N., Kirihata, T., & Iyer, S. S., (2013). A self-authenticating chip architecture using an intrinsic fingerprint of embedded DRAM. *IEEE J. Solid-State Circ., 48*(11), 2934–2943.

13. Wang, Y., Yu, W., Wu, S., Malysa, G., Suh, G. E., & Kan, E. C., (2012). Flash memory for ubiquitous hardware security functions: True random number generation and device fingerprints. In: *Proceedings of the IEEE Security and Privacy Symposium* (pp. 33–47).

14. Prabhu, P., Akel, A., Grupp, L. M., Yu, W. K. S., Suh, G. E., Kan, E., & Swanson, S., (2011). Extracting device fingerprints from flash memory by exploiting physical variations. In: *Proceedings of the International Conference on Trust and Trustworthy Computing* (pp. 188–2010).

15. Yuan, T., Buchanan, D., Wei, C., Frank, D., Ismail, K., Shih-Hsien, L., SaiHalasz, G., et al., (1997). CMOS scaling into the nanometer regime. *Proc. of IEEE, 85*(4), 486–504.

16. Orshansky, M., Nassif, S., & Boning, D., (2007). *Design for Manufacturability And Sialislical Design: A Constructive Approach*. Springer.
17. Pelgrom, M., Duinmaijer, A., & Welbers, A., (1989). Matching properties of MOS transistors. *IEEE Journal of Solid-Stale Circuits, 24*(5), 1433–1439.
18. Zhao, W., & Cao, Y., (2006). *New Generation of Predictive Technology Model for Sub-45nm Design Exploration* (pp. 7–12). ISQED.
19. Cao, Y., Sato, T., Orshansky, M., Sylvester, D., & Hu, C., (2000). *New Paradigm of Predictive MOSFET and Interconnect Modeling for Early Circuit Simulation* (pp. 201–204). CICC.
20. Cao, Y., (2012). *Predictive Technology Model.* Internet: http://ptm.asu.edu/ (accessed on 24 September 2021).
21. *International Technology Roadmap for Semiconductors (ITRS)*, (2015). Internet: https://www.semiconductors.org/resources/2015-international-technology-roadmap-for-semiconductors-itrs/ (accessed on 24 September 2021).
22. Maes, R., & Yerbauwhede, I., (2010). Physically unclonable functions: A study on the state of the art and future research directions. In: *Towards Hardware-Intrinsic Security* (pp. 3–37). Springer.
23. CHES, (2012). Emission Analysis of AES. In: *Cryptographic Hardware and Embedded Systems* (pp. 41, 57). Springer.
24. Skori_c, B., Tuyls, P., & Ophey, W., (2005). Robust key extraction from physical unclonable functions. In: *Applied Cryptography and Network Security* (pp. 407–422). Springer.
25. Skorobogatov, S., (2010). Optical fault-masking attacks. In: *Fault Diagnosis and Tolerance in Cryptography (FDTC)* (pp. 23–29). 2010 Workshop, IEEE.
26. Skorobogatov, S. P., & Anderson, R. J., (2003). Optical Fault Induction Attacks. In: *Cryptographic Hardware and Embedded Systems-CHES 2002* (pp. 2–12). Springer.
27. Tuyls, P., & Batina, L., (2006). RFID-tags for anti-counterfeiting. In: *Topics in Cryptology-CT-RSA 2006* (pp. 115–131). Springer.
28. Kim, K. H., Jo, S. H., Gaba, S., & Lu, W., (2010). Nanoscale resistive memory with intrinsic diode characteristics and long endurance. *Applied Physics Letters, 96*(5), 053106.
29. Koeberl, P., Kocaba_s, U., & Sadeghi, A. R., (2013). Memristor PUFs: A new generation of memory-based physically unclonable functions. In: *Proceedings of the Conference on Design, Automation and Test in Europe* (pp. 428–431). EDA Consortium.
30. Kvatinsky, S., Talisveyberg, K., Fliter, D., Friedman, E. G., Kolodny, A., & Weiser, U. C., (2011). *Verilog-a for Memristor Models*. Technical report, Citeseer.
31. Kwon, D. H., Kim, K. M., Jang, J. H., Jeon, J. M., Lee, M. H., Kim, G. H., Li, X. S., Park, G. S., Lee, B., Han, S., et al., (2010). Atomic structure of conducting nanofilaments inTiO_2 resistive switching memory. *Nature Nanotechnology, 5*(2), 148–153.
32. Linn, E., Rosezin, R., Kugeler, C., & Waser, R., (2012). Complementary resistive switches for passive nanocrossbar memories. *Nature Materials, 9*(5), 403–406.
33. Potkonjak, M., & Goudar, V., (2014). Public physical unclonable functions. *Proceedings of the IEEE, 102*(8), 1142–1156.
34. Taur, Y., & Ning, T., (1998). *Fundamentals of Modern VLSI Devices*. Cambridge University Press: Cambridge, UK.

35. Mondal, S., Talapatra, S., & Rahaman, H., (2011). Analysis, modeling and optimization of transmission gate delay. In: *Proceedings of the Asia Symposium on Quality Electronic Design (ASQED)* (pp. 246–253). Kuala Lumpur, Malaysia.

36. Katzenbeisser, S., Koçabas, Ü., Van, D. L. V., Sadeghi, A. R., Schrijen, G. J., Schröder, H., & Wachsmann, C., (2011). Recyclable PUFs: Logically reconfigurable PUFs. *J. Cryptogr. Eng., 1*, 177–186.

37. Gassend, B., Clarke, D., Van, D. M., & Devadas, S., (2002). Silicon physical random functions. In: *Proceedings of the ACM Conference on Computer and Communications Security* (pp. 148–160). Washington, DC, USA.

38. Maiti, A., & Schaumont, P., (2009). Improving the quality of a physical unclonable function using configurable ring oscillators. In: *Proceedings of the International Conference on Field Programmable Logic and Applications* (pp. 703–707). Prague, Czech Republic.

39. Ramazani, A., Biabani, S., & Hadidi, G., (2014). CMOS ring oscillator with combined delay stages. *Int. J. Electron. Commun., 68*, 515–519.

40. Joachims, T., (2003). Learning to classify text using support vector machines: Methods, theory and algorithms. *Comput. Linguist., 29*, 655–664.

41. Maiti, Y. G., & Schaumont, P., (2011). *A Systematic Method to Evaluate and Compare the Performance of Physical Unclonable Functions*. Cryptology e-Print Archive, Report 2011/657, Internet: http://eprint.iacr.org/ (accessed on 24 September 2021).

42. Schlosser, A., Nedospasov, D., Kramer, J., Orlic, S., & Seifert, J. P., (2012). Simple photonic emission analysis of AES. In: *Cryptographic Hardware and Embedded Systems-CHES 2012* (pp. 41–57). Springer.

43. Mahmoud, A., Ruhrmair, U., Majzoobi, M., & Koushanfar, F., (2013). *Combined Modeling and Side-Channel Attacks on Strong PUFs*. IACR Cryptology e-Print Archive, 2013, 632.

44. Maiti, V. G., & Schaumont, P., (2013). A systematic method to evaluate and compare the performance of physical unclonable functions. In: *Embedded Systems Design with FPGAs* (pp. 245–267). Springer.

CHAPTER 6

An Impact of Aging on Arbiter Physical Unclonable Functions

KURRA ANIL KUMAR and USHA RANI NELAKUDITI

Department of Electronics and Communications Engineering,
Vignan's Foundation for Science, Technology, and Research
(Deemed to be University), Guntur, Andhra Pradesh, India,
E-mail: kakumar94@gmail.com (K. A. Kumar)

ABSTRACT

On chip physical unclonable functions (PUFs) are emerged to be one of the lightweight security primitives in hardware cryptography to solve several security problems. It can generate a random set of responses for each challenge by utilizing the inherent process variations during manufacturing process. The extracted responses from each PUF should be die specific and acts like a device fingerprint. Hence, it is difficult to clone or counterfeit the similar kind PUF architecture. It is well known that the responses obtained from each PUF can be utilized for the different kinds of cryptographic applications, including IP protection, device authentication, RFID detection/identification, IoT security, etc. The performance of PUF architectures is estimated by using the metrics like uniqueness, reliability, and randomness. A PUF should be robust against temporal variations in circuits and the effect of temporal changes in circuits are influences the reliability. Out of the several factors, aging is one of the major and critical factors which flip the PUF response. In practice, it is essential to identify the major factors responsible for aging such as bias temperature instability (BTI), hot carrier injection (HCI), time-dependent dielectric breakdown (TDDB), electromigration, etc. This chapter mainly addresses the impact of aging on strong PUFs and estimated the aging phenomenon on MUX-decoder PUF architecture by applying statistical analysis using cadence virtuoso 45 nm CMOS technology.

6.1   INTRODUCTION

As the day-by-day technology has shrunk down as a result fabrication cost of the Integrated circuits (ICs) has increased enormously, hence counterfeiting of ICs becomes one of the critical issues facing by semi-conductor manufacturing companies. Many methods have been employed to identify and authenticate these issues, like secure split test (SST), hardware metering, combating die IC recovery, physical unclonable functions (PUFs), etc. Out of the several counterfeiting mechanisms, PUFs are emerged be a one of the dominant low costs, lightweight, and low area security primitive in hardware cryptography. Gassend et al. [1, 2] are introduced the concept of silicon PUFs by leveraging the intrinsic properties of ICs, and PUFs are die-specific random functions, which incurs generates a specific set of response (Ri) for each set of challenges (Ci) by utilizing the unpredictable and unavoidable tiny process variations during the fabrication process [3]. Due to these fabrication variations makes each IC will acts as a die specific fingerprint. Figure 6.1 illustrates the primary CRP mechanism in each PUF. The basic operation of any PUF can be characterized by considering the CRP mechanism. The responses generated from the PUF are mainly depended on two types of mismatches such as technology-dependent and technology independent. The parameters such as oxide thickness, masking, doping concentration, channel length, geometry of transistors, etc., are mainly responsible for technology-dependent, and on the other hand, voltage, temperature, and noise significantly impact local mismatches [4, 5]. Figure 6.2 depicts the various mismatches occurring during the manufacturing process of an IC. Depending upon the number of challenge-response pairs (CRPs), PUFs can be explicitly distinguished into two types, such as strong PUFs and weak PUFs.

FIGURE 6.1   Challenge-response mechanism in PUFs.

Strong PUFs are a class of PUFs having an exponential number of CRPs. Therefore the adversary cannot enumerate total set of CRPs from a PUF in a fixed amount of time. The obtained responses from PUFs are highly stable over a wide range of operating conditions (like temperature and voltage) and can withstand several numbers of attacks. However, a strong PUFs does not require any extra set of hardware to authentication as well as its responses are highly unique and quite unpredictable. It also noted that the mathematical modeling for strong PUFs is different from weak PUFs. Pappu et al. was proposed first strong PUF such as optical PUF [6–8]. It consists of an optical scattering object, assumed to be a plastic token, when a set of a laser beam is directed towards, this token scatters with an angle, and the resulting response will be elevated into different dimensions. For each optical beam will approximately create 1010 number of independent CRPs. Figure 6.3 depicts the basic architecture of the optical PUF. And several sets of electrical PUFs are being implemented based on electrical characteristics like arbiter PUFs, ring oscillator PUF, etc.

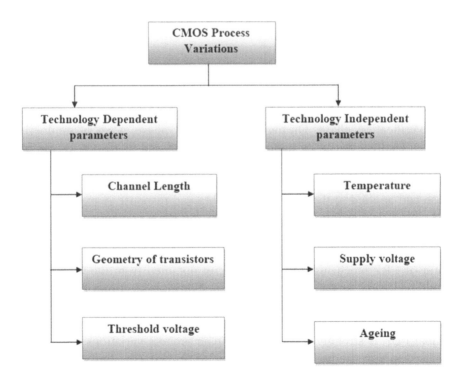

FIGURE 6.2 Mismatch mechanisms in PUFs.

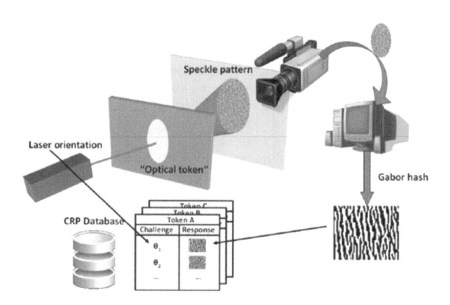

FIGURE 6.3   Architecture of the optical PUF.

The strong PUFs can be mainly designed for device authentication purposes without any requirement of the cryptographic module; this process can be carried out in two phases, such as the enrollment phase and verification phase. During the enrollment phase, a set of randomly chosen challenges can be applied to the device, and its responses can be stored in the database for future authentication. And while the verification stage, a trusted party chose a set of challenges from the recorded database and applied to the device, if the obtained response is matched with the stored response in the database, then the IC is authenticated [8–11]. The second category of the PUFs is the weak PUFs, which are having a limited fixed number of CRPs and exhibits a physical disorder in the form of CRP mechanism [12]. The secret obtained from the weak PUFs could be treated as the physical obfuscated keys. Like strong PUFs, weak PUFs are also robust against the environmental variations and withstand over a wide range of the voltages. The response generated from the individual PUF is strongly relying on intrinsic process variations. These can be used in very limited applications due to fixed/linear number of CRPs, and if the secret can prevail, it can be easily modeled by machine learning (ML) attacks and invasive attacks [13–15]. Therefore, there are several numbers of weak PUFs are there, such as SRAM PUF, butterfly PUF,

Anderson PUF, etc. The usage of PUFs has been demonstrated from device authentication to identification applications [16]. In the current scenario, PUFs mainly suffer from two critical issues, such as instability of PUF responses and ML-based "modeling building attacks." An adversary can duplicate the PUF architectures by measuring the delays of the PUF using differential power analysis and estimating the statistical variations by direct measurement by opening up the package. The adversary can able to build an exact model by measuring the responses by arbitrarily applying challenges. Moreover, PUFs suffer from various types of attacks such as brute-force attacks, modeling attacks, optical emission attacks, timing, and power attacks, respectively. On the other hand, reliability is also one of the significant issues in the device authentication process. The responses generated from each PUF are mainly dependent on process variations, but various uncontrollable factors such as environmental noise, temperature, and aging degrade the performance of the PUF metrics. The effect of environmental noise mainly occurs when operating the circuits at different operating conditions like temperature and voltage. In contrast to environmental noise aging is also one of the primary phenomena, which impact the reliability of the PUF. This is mainly caused due to undesirable changes in the physical structure of the device due to negative bias temperature instability (NBTI), hot carrier injection (HCI), time-dependent dielectric breaks down (TDDB), and electromigration. This chapter mainly focuses on the impact of aging on the strong PUFs, more specifically arbiter PUFs. The rest of the chapter is organized as follows. Section 6.1 explains the introduction to the PUFs and major classifications of PUFs. Section 6.2 describes the different security metrics of the PUF. Section 6.3 indicates the various mechanisms to degrade the reliability of the PUF, and Section 6.4 briefs the impact of the aging on the reliability of strong PUFs and estimation of aging for proposed mux-decoder PUF. Finally, Section 6.5 concludes the chapter.

6.2   METRICS OF PUF

Here we are introducing a set of metrics to evaluate the performance of PUF design and, more specifically, which PUF is suitable for which application [16–18]. Therefore, we were introducing the three different metrics such as uniqueness, reliability, and uniformity.

6.2.1 UNIQUENESS

It is a measure of the ability of PUF to generate a unique set of responses for each set of the challenge. The number of bits could distinguish the behavior of the PUF has unique when the same challenge is applied to the two PUFs from the same wafer. The ideal value of any PUF is 50%, and its mean value is varying from 20% to 50%. In practice, the performance of PUF is based on Interchip variations and body bias effect, respectively. Therefore the average Interchip hamming distance for 'K' PUFs can be described by Eqn. (6.1):

$$\text{Uniquness} = \frac{2}{K(K-1)} \sum_{K}^{K-1} \sum_{j=i+1}^{K} \frac{HD(R_i, R_j)}{N} \times 100 \qquad (6.1)$$

Let R_i and R_j be n-bit responses of two different chips i and j, to the same input challenge 'C,' 'N' indicates the total number of PUF instances and hamming distance (HD), respectively. Figure 6.4 illustrates evaluation of uniqueness for two different PUFs with the same challenge. From Figure 6.4, it clearly emphasizes the same challenge (0010101) being applied to the two PUFs (PUF1 and PUF2) at ambient temperature (27°C) and obtained response is differs the two bits of hamming distance with each other, respectively.

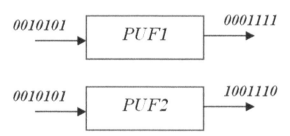

FIGURE 6.4   Uniqueness evaluation of a PUF design.

6.2.2 RELIABILITY

It is the ability to measure the stability of the PUF response (R) for a corresponding challenge (C) over a different set of environmental conditions such as voltage, temperature, and aging, etc. It can be evaluated by "intra-chip hamming distance (HDintra)." Mathematically it can be expressed by Eqn. (6.2). Ri(n) and Ri,t(n) represents the n-bit response at ambient temperature

(Ri) and Ri,t(n) n-bit response at different operating conditions for same challenge. 'S' indicates a number of chips for N bit PUF.

$$HDintra = \frac{1}{S}\sum_{t=1}^{s} \frac{HD\left(R_i\left(n\right), R_{i,t}\left(n\right)\right)}{N} \times 100 \qquad (6.2)$$

The overall reliability of any PUF can be represented by Eqn. (6.3):

$$Reliability = 100\% - HDintra \qquad (6.3)$$

In order to estimate the reliability applied the same challenge (0010101) for PUF1 at different operating conditions such as 27°C and 57°C temperatures (As shown in Figure 6.5).

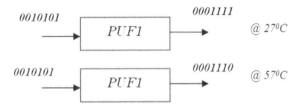

FIGURE 6.5   Reliability evaluation of a PUF design.

6.2.3   UNIFORMITY

It is the estimation of the unpredictability of the PUF response even prior knowledge of the circuit. It defines the total number of the "0" and "1" in a given set of the response, and the ideal value of the uniformity for PUF could be 50%. Mathematically, it could be expressed by Eqn. (6.4).

$$Uniformity = \frac{1}{n}\sum_{i=1}^{n} R_{i,l} \times 100\% \qquad (6.4)$$

where; 'n' indicates the total number of responses; Ri indicates the hamming weight of the ith response.

6.3   RELATED WORK

Reliability is one of the critical metrics, and the designer must take consideration while estimating the stability of any particular PUF design. Especially in cryptographic applications and device authentication, reliability places a vital role. There are some sets of mechanisms which are highly affects the reliability of the PUF include temperature, power supply, noise, aging effects, etc.

6.3.1   EFFECT OF TEMPERATURE ON RELIABILITY OF THE PUF

Temperature can affect the switching activity of the circuitry when sudden increases or decreases and (uncontrollable abnormalities) it tends to changes the behavior of the circuit, which eventually leads to hardware malfunction [19, 20]. Due to these abnormalities potentially, the delays of transistors can change in a random manner. Hence the overall response of the PUF can vary in accordance with the corresponding challenge. Therefore, it affects the authentication protocol.

6.3.2   AGING MECHANISMS ON RELIABILITY OF THE PUF

Aging is an undesirable process for any digital circuits; it merely affects the IC and deteriorates the performance by means of leakage current, more power consumption. Hence this causes functional failures. As long as prolonged use of devices, the delays of the circuits tend to drift and if changes are quite high, then it might be difficult to recognize a PUF undergone to measure. Several aging mechanisms can impact on the reliability of the IC during its whole lifetime. The behavior of the digital circuits mainly depends on current and voltage characteristics of the transistors interconnect capacitance, etc., as long as devices get aging its influences the I-V characteristics and effects parameters like delay, speed. The primary sources for aging of any ICs are mainly due to NBTI, TDDB, HCI, electromigration, respectively. These mechanisms can be described clearly in further subsections.

6.3.3   NEGATIVE BIAS TEMPERATURE INSTABILITY (NBTI)

NBTI is one of the major factors affecting the performance of the PMOS in ICs. Usually, when a negative voltage is applied to PMOS transistors gate. In this case, positive interface traps are created over the si-sio2 interface and also trapping of holes in the dielectric bulk [21, 22]. Hence, this effect increases the threshold voltage of the PMOS transistors abruptly and retards the drain current (Id) by slowing down the circuit. Depending upon the bias condition of PMOS, transistors exhibit the NBTI in two phases, such as the stress phase and recovery phase (as shown in Figures 6.6(a), (b), and (c)). During the stress phase, the negative voltage is applied to its gate then transistor becomes (Vgs < Vt, threshold voltage Vt being negative for PMOS) ON. While recovery phase, a positive voltage is applied to the gate of PMOS transistor, during this phase threshold. Figure 6.7 depicts the threshold voltage drift of the PMOS transistor under the stress and recovery phase over 6 months [23].

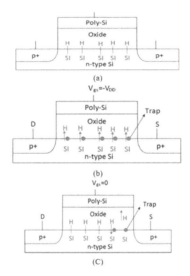

FIGURE 6.6 (a) NBTI effect for fresh PMOS; (b) NBTI effect for stress phase; and (c) recovery phase.

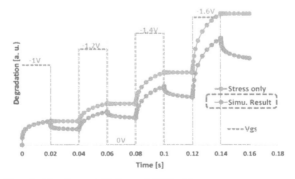

FIGURE 6.7 Threshold voltage shift due to NBTI effect.

When stress is removed, then the device is turned off. In the same way, NMOS transistors get affected by its performance due to positive bias temperature instability (PBTI). But the performance degradation is less in less prominent PBTI as compared to the NBTI. The amount of threshold voltage shift can be expressed by Eqn. (6.5).

$$\Delta V_t = Ke^{\frac{E_{ox}}{E_0}} t^{0.25} \tag{6.5}$$

6.3.4 HOT CARRIER INJECTION (HCI)

Amid the switching of a transistor, some of the high energy charge carriers get injected into the gate oxide. HCI is a major unwavering quality issue for NMOS as electrons have higher mobility than holes [24, 25]. The amount of HCI-related aging degradation of a circuit is directly dependent on the activity of the circuit. Circuits working at higher VDD hoisted temperatures, and high switching activity is exceptionally vulnerable to HCI-related aging. This phenomenon could occur when transistors are in pinch-off condition, and unlike BTI, the amount of traps occurred in HCI are permanent. Figure 6.8 shows the HCI mechanism in the NMOS transistors.

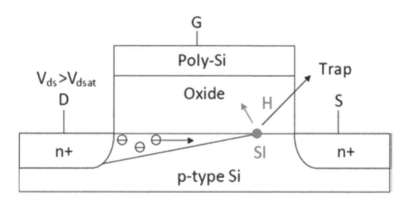

FIGURE 6.8 Effect of HCI aging in NMOS transistor.

This mechanism can be effectively characterized by shifting the Vth of the NMOS device and which is proportional to the number of the traps, respectively.

$$\Delta N_{it}(t) = C[\frac{I_{ds}}{W} \ e^{\left(\frac{\varnothing_{it,e}}{q\lambda_e E}\right)}t]^n \qquad (6.6)$$

where; C is the technology-dependent constant; Ids is the drain to source current; W is the width of the transistor; $\varnothing_{it,e}$ is the energy required to overcome the oxide barrier; E is the electrical field at the drain; λ_e is the hot electron mean free path; t is the operating time during which transistor under stress condition.

6.3.5   TIME-DEPENDENT DIELECTRIC BREAKDOWN (TDDB)

When the thickness of devices created within the dielectric is quite high, it can make the dielectric to breakdown, bringing about a conductive way over the gate oxide. This is an issue in the technologies as supply voltages, and the related electric fields have not been scaling as forcefully as the device dimensions. Since of these, a little thickness of traps is adequate for the oxide to breakdown. As a result, the impact of TDDB has turned out to be less extreme [26]. The degradation of the performance of the CMOS devices eventually affects the metrics of the PUF circuits. BTI is one of the most likely occurred mechanisms in weak PUFs such as SRAM PUFs, butterfly PUFs, etc. On the other hand, mechanisms such as electromigration and hot electron injection affect the performance of the ring oscillator PUF. Most of the failures occur when the transistor parameters are operated beyond certain tolerance limits, and this creates a negative impact on the circuit performance. Over the last decade, a lot of research has been done to eliminate the different degradation mechanisms during the manufacturing and design phase. Many alternative techniques have been deployed to bypass faults and to minimize the degradations, such as adding redundant hardware, clock, and power gating, etc. Degradation mechanisms like TDDB, HCI, and NBTI are mainly dependent on frequency, temperature, and voltage stress.

The process of estimating the aging effect is carried out in two ways, such as inducing the degradation model and integrating the model into a simulation tool. The most commonly used tools, such as cadence relXpert and synopsis H-spice MOSRA, are extensively used to estimate the performance metric (reliability) of the electrical circuit. This process can be carried out in two stages, such as the pre-stress and post-stress stage, respectively. During the pre-stress case, we can measure the Worst-case behavior of the circuit can be affected by the NBTI, HCI, and TDDB and secondary metrics of the circuit such as delay, leakage power can be by effectively measured by conducting the post-stress analysis. The evaluation methodology of the aging can be done by four stages, as shown in Figure 6.9.

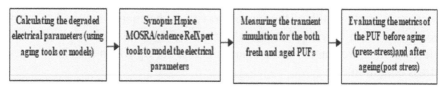

FIGURE 6.9   Evaluation diagram for impact of aging on silicon PUFs.

Flowchart 9 represents the steps involved during the evaluation of the aging under different conditions using HSPICE. An input technology library and operating temperatures are predefined. In the first phase, we need to estimate the performance of the circuit under zero timing conditions (no aging). Then, using HSPICE MOSRA to evaluate the performance of the circuit at different stages under different operating conditions. This process is carried out in two phases, such as pre-stress and post-stress phases, respectively. During the pre-stress phase, inputs are given at different environmental conditions and estimate the threshold voltage of the MOSFETs along with its electrical parameters at different conditions, and various intervals of time produces the report based on MOSRA models. The next step involves calculating the behavior of the circuit at post-stress analysis. During this phase, the percentage of characteristics of the circuit could differ for the actual circuit that could be analyzed (Figure 6.10).

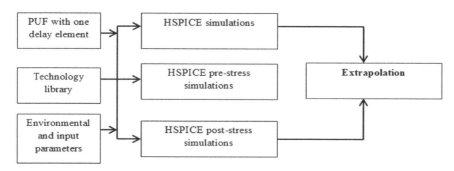

FIGURE 6.10   Flowchart for extracting the delay parameters of a delay-PUF with an arbitrary size.

6.3.6   SOURCES OF SOFT ERRORS IN CMOS ICS

Soft errors are another category of the errors that usually occur in CMOS semiconductor technologies. Basically, there are two types of soft errors, such as transient faults (occur mainly due to environmental conditions) and intermittent failures (non-environmental conditions). In this, we are going to discuss the different types of primary sources that are responsible for soft errors such as radiation-induced errors, crosstalk noise, ground bounce, and thermal noise, respectively.

6.3.7   RADIATION-INDUCED SOFT ERRORS

These errors mainly occur due to two primary sources, such as neutrons, cosmic rays, and alpha particles. This could mostly occur when an electronic device operated at very high altitudes, such as aerospace and avionic applications. The radiations are directed to the combinational logic, which may induce the error in the transient pulse. This phenomenon mainly occurs in memory devices and sequential circuits, respectively. The momentum when the radiation hits the combinational circuit could affect the transients, and this affects the carries on a long chain of the logic. There are several techniques that have been employed to mitigate the soft errors, and radiation methods are used to protect the storage elements (Figure 6.11).

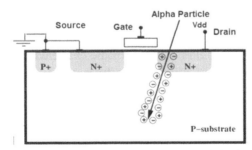

FIGURE 6.11   The generation of free carriers by an alpha particle hit on an NMOS device.

6.3.8   CROSS TALK NOISE

This is another dominant type of reliability issue, and it is mainly classified into two types, such as functional failures due to static noise and timing failures due to an increase in delay of the communication channel link. This phenomenon impacts the functionality of the device and forces it to works as a malfunction, mainly due to the capacitive coupling. The crosstalk effect should be permanent, and this effect should be excessive as a sudden increase of the overvoltage at gate input of the device.

6.3.9   GROUND BOUNCE

This phenomenon usually happened when the unexpected rise of the current drawn from the voltage at the rising edge of the clock in synchronous systems. In such a case, the voltage may spike at an abnormal level; these glitches effectively lower the speed of the device and which is directly proportional to the applied voltage across the network of the system.

6.4   ARBITER PUFS

6.4.1   ARCHITECTURE AND OPERATION OF ARBITER PUFS

In short, the underlying architecture of the arbiter PUF, as shown in Figure 6.12; it consists of two stages, such as the delay network and arbiter network. The delay network consists of a cascaded connection of multiple delay stages. Each delay stage consists of two multiplexers, which are connected at the top and bottom stage, respectively. The selection for the multiplexer will act as the challenge bit at every stage, therefore depending upon several applied challenges and due to random process variations will make a difference in the delay stages. For each set of the challenge, it will propagate to either the cross or straight. Finally, the arbiter is fed at the final stage of the delay network, which translates the analog timing delay difference into the corresponding digital value. Usually, SR latch or D flip-flop is used as the arbiter to detect the timings delay differences accurately [27–30].

Like arbiter PUFs, several types of PUFs have been proposed to enhance the metrics of the PUF (uniqueness, reliability, and uniformity). The classical arbiter PUF suffers from the linear timing delay difference, which makes it easy to predict the final response of the PUF. Hence to overcome the above-mentioned problem and to enhance the unpredictability proposed a set of non-linear PUFs, such as feed-forward mux PUF, feed-forward mux PUF overlap structure, feed-forward mux PUF cascade structure, feed-forward mux PUF separate structure, etc. This architecture includes a set of arbiters placed in the middle of the PUF structure; the output generated from each arbiter will acts as the challenge for the further stages, respectively. This architecture significantly improves the numerical modeling attacks. However, the reliability of the PUF has been easily degraded due to environmental effects. The notations that can be used in this chapter can be explained in Table 6.1.

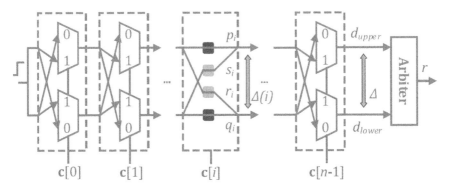

FIGURE 6.12   Basic architecture of the silicon arbiter PUF.

TABLE 6.1   Notations Used in This Chapter

Notation	Explanation
N	Number of mux stages
C	The number of applied challenge vectors
A	Input excitation for the delay network
R	Response of the final PUF
S	Number of stages in the PUF
D^i_{top}	Delay of the top multiplexer at i^{th} stage
D^i_{bottom}	Delay of the bottom multiplexer at i^{th} stage
Δ_i	Delay difference between top and bottom elements at the ith stages
R_n	Delay difference of N stages

6.4.2   MATHEMATICAL MODELING OF ARBITER PUF

In a real time scenario, the amount of process variations that occurred at each and every stage can be modeled by using statistical static timing analysis (SSTA). In the same way, the amount of delay variations can be approximated by using the Gaussian distribution. Based on the above discussions we can model each multiplexer as an independent random variable Di, represented by the Gaussian random variable such as N(μ, σ^2), where μ represents the mean and σ^2 indicates the standard deviation of delay of the mux. The delay difference between top and bottom MUXes can be represented by $\Delta_i = D_i^t - D_i^b \sim N(0, 2\sigma^2)$. the final response obtained from the last stage can be modeled by $R_N = \sum_{j=1}^{N}(-1)^{Ci'}\Delta_i$ where $Ci' = \phi_{j=i+1}^{N}C_j$ and $c_N' = 0$. The final response generated from the final stage can be given by R = sign(R_N) = 1 *if Rn* is greater than 0 else output is 0, respectively.

6.4.3 ESTIMATION OF AGING ON ARBITER PUFS

It is an undesirable effect, and it changes the functionality of the hardware device. Therefore, it produces an unavoidable response. Hence eventually, it could increase or decrease the delay of the circuit. NBTI, HCI, and TTDB are the significant factors for contributing to the reliability of the PUF, and the effect of the aging on strong PUFs such as arbiter PUFs can be described adequately. The basic architecture of the arbiter PUF consists of the two stages, such as delay network stage and arbiter stage. The amount of delay variations occurred at each stage can be estimated from the Gaussian distribution curve by measuring the mean and standard deviation, respectively. The conventional arbiter PUF multiplexer will acts as a switching element in the delay network. The variability of the delay can be varied either top mux or bottom mux due to aging effect. The delay associated with the top and bottom multiplexer can be explicitly represented by D^i_{top} and D^i_{bottom}, respectively and its delay difference can be expressed by $\Delta^i = D^i_{top} - D^i_{bottom}$. The Gaussian distribution of the delay can be attributed in terms of mean and standard deviation N (μ, σ^2). The delay difference of the ith stage can be expressed by N (0, $2\sigma^2$). Due to the gradual aging of the device it can lead to delay variability which can be modeled by N (μ', σ_a^2) [37–45]. The final delay difference of the aged PUF at ith stage can be represented by Eqn. (6.7):

$$\Delta^i_{aged} = \Delta^i \left(1 + \frac{\Delta^i_{aged} - \Delta^i}{\Delta^i}\right) = \Delta^i (1 + p_i)$$

(6.7)

where; p_1 is the percentage of the delay difference in the ith stage. Therefore, the Gaussian distribution can be used to model the randomness of the PUF. Figures 6.13(a) and (b) illustrates the variation on delay-difference represented in terms of samples over 2 and 6 hours, respectively.

6.4.4   AGING MODEL FOR ARBITER PUF

The arbiter is one of the final and fundamental blocks which decides the final response depending upon the arrival of the two signals. The delay at the final stage can be modeled based on the propagation delay. The delay difference of the aged arbiter can be expressed by Eqn. (6.8):

$$\Delta_{agd}^{arb} = \Delta^{arb} \left(1 + p_i\right)$$ (6.8)

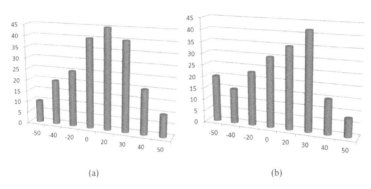

(a)   (b)

FIGURE 6.13   Variation of the delay differences during 2 and 6 hours.

As long the device is gets aged the amount of bit error rate can be estimated over the 20 months and in the same way the variation of the delay with respect to mean and standard deviation observed. The amount of delay observed from each delay network can be measured with respect to mean and standard deviations over 0 to 20 months corresponding challenge oX5A5A.

6.4.5   ARCHITECTURE OF THE MUX-DECODER PUF ARCHITECTURE

As shown in Figure 6.14, the architecture of the 8-stage mux decoder PUF is similar to the conventional arbiter PUF. Each stage in the delay network consists of one decoder and two multiplexers. A pulse signal is given as input for the decoder and a unique challenge is given to the top and bottom multiplexers at a time, depending upon input challenge and applied input the data is transferred to the one stage to other stage, respectively. The final stage of the delay network is connected to the arbiter, which decides the final response depends upon the arrival of the signal to the arbiter (SR-latch).

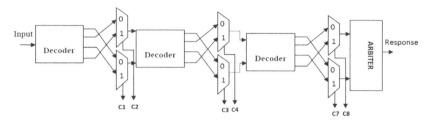

FIGURE 6.14 Basic architecture of the mux-decoder arbiter PUF.

6.4.6 SIMULATION SETUP AND STATISTICAL PERFORMANCE ANALYSIS OF PROPOSED PUF

The response of the PUF design is mainly relayed on the minute fabrication process variations, to estimate the amount of process variations with respect to delay performed the statistical analysis. In the proposed circuit was simulated using cadence virtuoso 45 nm technology. Estimated the process variations by applying Monte-Carlo (MC) analysis and observed its output response from the six-sigma plot. The response of the PUF design can be measured with respect to its security metrics such as uniqueness, reliability, and randomness, respectively. While calculating the response of the PUF circuit chosen 50 sets of randomly challenges performed the 200 iterations at each stage by applying 200 number of samples at each corner (slow-fast, fast-fast (FF), slow-slow, fast-slow). The uniqueness of the PUF circuit were measured at five different instances, and supply voltage of 0.45 V at 27°C temperature and its corresponding Interchip hamming distance should be 47.51%, 47.63%, 47.82%, 48.21%, and 48.36% respectively. Similarly, we measured the stability of the PUF circuit at different operating conditions by varying temperatures from 27°C to 80°C. Observed the amount of deviation of rising time, falling time, and delay at four different corners. The overall reliability of the PUF circuit was measured by applying a soft response technique. To evaluate its performance chosen to be 0.45 v and 27°C are the reference value and varied the voltage ranging from 0.45 v to 0.9 v, respectively. The maximum reliability of the PUF circuit could be 88% at 27°C and 49.36% at 0.45 v, respectively.

As shown in Figures 6.15(a) and (b), the number of delay variations concerning the aging could be observed over two hours and four hours. As the device gets, an aging delay of the circuit could increase linearly. Therefore the amount of occurrence of the bit error rate could be increased enormously; hence, it could affect the overall performance of the circuit.

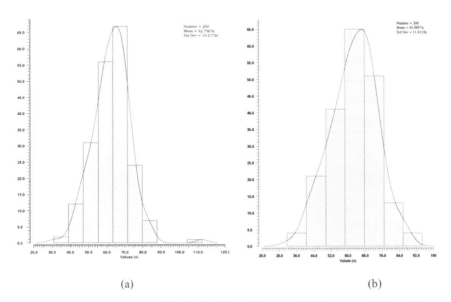

(a)   (b)

FIGURE 6.15   Percentage of delay distribution at 2 hours and 4 hours over 200 samples.

6.5   CONCLUSION AND FUTURE SCOPE

We have characterized the notation of the physical random/unclonable functions (PUFs) and significantly illustrated the application of the importance of the PUF during the process of authentication. In this chapter, we investigated the significant mechanisms contributing to the effect of aging. We analyzed the impact of the aging on the reliability of the strong PUFs, such as the arbiter PUFs. Analyzed the impact of the switching activity on each delay element and arbiter with respect to mathematical modeling. Therefore the effect of the aging will predominately impact on the targeted application. The simulation results describe the impact of the delay of each stage of the switching element and arbiter.

Fabrication of proposed and thoroughly tested decoder-mux based strong PUF architecture can be considered for future work since it can be used in many real-time IoT authentication applications.

KEYWORDS

- **challenge-response pair**
- **hardware security**
- **hot carrier injection**
- **modelling attacks**
- **Monte-Carlo analysis**
- **physical unclonable functions (PUFs)**
- **time-dependent dielectric breakdown**

REFERENCES

1. Alam, M. A., Kufluoglu, H., Varghese, D., & Mahapatra, S., (2007). A comprehensive model for PMOS NBTI degradation: Recent progress. *Micro Electron Reliability, 47*(6), 853–862.

2. Bhargava, M., & Mai, K., (2014). An efficient reliable PUF-based cryptographic key generator in 65nm CMOS. In: *Design, automation & test in Europe (DATE)* (pp. 70:1–70:6).

3. Cao, Y., Zhang, L., Chang, C., & Chen, S., (2015). A low-power hybrid RO PUF with improved thermal stability for lightweight applications. *IEEE Trans CAD Integrated Circuits Syst 34*(7), 1143–1147.

4. Cha, S., Chen, C. C., Liu, T., & Milor, L. S., (2014). Extraction of threshold voltage degradation modelling due to negative bias temperature instability in circuits with I/O measurements. In: *VLSI test symposium (VTS)* (pp. 1–6).

5. Cherif, Z., Danger, J., Guilley, S., & Bossuet, L., (2012). An easy-to-design PUF based on a single oscillator: The loop PUF. In: *Digital System Design (DSD)* (pp. 156–162).

6. Pelgrom, M. J., Duinmaijer, A. C., & Welbers, A. P., (1989). Matching properties of MOS transistors. *IEEE Journal of Solid State Circuits, 24*(5), 1433–1439.

7. Holcomb, D. E., Burleson, W. P., & Fu, K., (2009). Power-up SRAM state as an identifying fingerprint and source of true random numbers. *IEEE Trans. on Computers, 58*(9), 1198–1210.

8. Morozov, S., Maiti, A., & Schaumont, P., (2010). An analysis of delay based PUF implementations on FPGA. In: *Reconfigurable Computing: Architectures, Tools and Applications* (ARC) (pp. 382–387).

9. Kuuoglu, H., & Alam, M. A., (2007). A generalized reaction-diffusion model with explicit h-h2 dynamics for negative-bias temperature-instability (NBTI) degradation. *IEEE Trans. on Electron Devices, 54*(5), 1101–1107.

10. Lu, Y., Shang, L., Zhou, H., Zhu, H., Yang, F., & Zeng, X., (2009). Statistical reliability analysis under process variation and aging effects. In: *Design Automation Conference (DAC)* (pp. 514–519).

11. Chakravarthi, S., Krishnan, A., Reddy, V., Machala, C. F., & Krishnan, S., (2004). A comprehensive framework for predictive modeling of negative bias temperature instability. In: *Reliability Physics Symp.* (pp. 273–282).

12. Saha, D., Varghese, D., & Mahapatra, S., (2006). Role of anode hole injection and valence band hole tunnelling on interface trap generation during hot carrier injection stress. *IEEE Electron Device Letters, 27*(7), 585–587.

13. Rodriguez, R., Stathis, J., & Linder, B., (2003). Modeling and experimental verification of the effect of gate oxide breakdown on CMOS inverters. In: *IEEE Int'l Reliability Physics Symposium* (pp. 11–16).

14. Sinanoglu, O., Karimi, N., Rajendran, J., Karri, R., Jin, Y., Huang, K., & Makris, Y., (2013). Reconciling the IC test and security dichotomy. In: *European Test Symp. (ETS),* (pp. 1–6).

15. Khan, S., Haron, N. Z., Hamdioui, S., & Catthoor, F., (2011). NBTI monitoring and design for reliability in nanoscale circuits. In: *Int'l Symp. on Defect and Fault Tolerance in VLSI and Nanotechnology Systems (DFT)* (pp. 68–76).

16. Karimi, N., Danger, J. L., Guilley, S., & Lozach, F., (2016). Predictive aging of reliability of two delay PUFs. In: *Security, Privacy, and Applied Cryptography Engineering (SPACE)* (pp. 213–232).

17. Karimi, N., Danger J., Slimani, M., & Guilley, S., (2017). Impact of the switching activity on the aging of delay-PUFs. In: *European Test Symp. (ETS)*.

18. Karimi, N., Guilley, S., & Danger, J. L., (2018). Impact of aging on template attacks. In: *Proceedings of the ACM Great lakes symposium on VLSI (GlSVLSI)* (pp. 455–458).

19. Zhou, C., Satapathy, S., Lao, Y., Parhi, K. K., & Kim, C. H., (2016). Soft response generation and thresholding strategies for linear and feed-forward MUX PUFs. In: *Proceedings of the 2016 International Symposium on Low Power Electronics and Design* (pp. 124–129). ACM.

20. Avvaru, S. S., Zhou, C., Satapathy, S., Lao, Y., Kim, C. H., & Parhi, K. K., (2016). Estimating delay differences of arbiter PUFs using silicon data. In: *2016 Design, Automation & Test in Europe Conference & Exhibition (DATE)* (pp. 543–546). IEEE.

21. Liu, W., Yu, Y., Wang, C., Cui, Y., & O'Neill, M., (2015). RO PUF design in FPGAs with new comparison strategies. In: *Circuits and Systems (ISCAS), 2015 IEEE International Symposium on* (pp. 77–80). IEEE.

22. Xu, X., Burleson, W., & Holcomb, D. E., (2016). Using statistical models to improve the reliability of delay-based PUFs. In: *2016 IEEE Computer Society Annual Symposium on VLSI (ISVLSI)* (pp. 547–552). IEEE.

23. Marsaglia, G., (2005). Ratios of normal variables and ratios of sums of uniform variables. *Journal of the American Statistical Association, 60*(309), 193–204.

24. Lao, Y., Tang, Q., Kim, C. H., & Parhi, K. K., (2016). Beat frequency detector based high-speed true random number generators: Statistical modelling and analysis. *ACM Journal on Emerging Technologies in Computing Systems (JETC), 13*(1), 9.

25. Lim, D., Lee, J. W., Gassend, B., Suh, G. E., Van, D. M., & Devadas, S., (2005). Extracting secret keys from integrated circuits. *IEEE Transactions on Very Large Scale Integration (VLSI) Systems, 13*(10), 1200–1205.

26. Sahu, C., & Singh, J., (2015). Potential benefits and sensitivity analysis of dopingless transistor for low power applications. *IEEE Transactions on Electron Devices, 62*(3), 729–735.

27. Shrivastava, V., Kumar, A., Sahu, C., & Singh, J., (2016). Temperature sensitivity analysis of dopingless charge-plasma transistor. *Solid-State Electronics, 117*, 94–99.

28. Yanambaka, V. P., Mohanty, S. P., Kougianos, E., Sundaravadivel, P., & Singh, J., (2017). Dopingless transistor based hybrid oscillator arbiter physical unclonable function. In: *Proceedings of the IEEE Computer Society Annual Symposium on VLSI (ISVLSI).*

29. Rahman, M. T., Forte, D., Fahrny, J., & Tehranipoor, M., (2014). ARO-PUF: An aging-resistant ring oscillator PUF design. In: *Proceedings of the Design, Automation Test in Europe Conference Exhibition (DATE)* (pp. 1–6).

30. Ruhrmair, U., Busch, H., & Katzenbeisser, S., (2010). Strong PUFs: Models, constructions, and security proofs. In: *Towards Hardware-Intrinsic Security* (pp. 79–96). Berlin, Germany: Springer.

CHAPTER 7

Advanced Power Management Methodology for SoCs Using UPF

USHA RANI NELAKUDITI, NAVEEN KUMAR CHALLA, and
KURRA ANIL KUMAR

*Department of Electronics and Communications Engineering,
Vignan's Foundation for Science, Technology, and Research
(Deemed to be University), Guntur, Andhra Pradesh – 522213, India,
E-mail: kakumar94@gmail.com (K. A. Kumar)*

ABSTRACT

Recently emerged portable high-performance battery-powered medical equipments, MP3 and MP4 music systems, cell phones, high resolution digital cameras, camcorders, and high-definition TV (HDTV) sets operate at low power to avoid frequent recharge and ensure long battery life. Microprocessors and DSP processors operating at GHz frequencies and high packaging density mobile devices with many functions undergo large-signal switching transitions leads to power consumption. The dissipated power will be appeared as a heat and leads to the temperature rise, causes the malfunction of semiconductor devices, affects the reliability. Hence, fans and cooling mechanisms are necessary to reduce the dissipated heat and ensure the reliability as well as reduction of design cost. The above-stated challenges demand the semiconductor devices operation at low voltage and low power. Hence, at present VLSI design was transformed to three-dimensional space, with power also as a design parameter along with chip density, performance. This chapter dealt with the low power techniques and power aware methods are considered to meet the requirements in SoCs. In VLSI, implementation can be performed at various abstraction levels. Similarly, low power methodologies incorporated in VLSI can also span to various abstraction levels from device to system. Based on the level of implementation design parameters will vary. For example, at device level, device, and interconnects geometry, threshold voltage (Vt) are the parameters to design low power methods. Clocking strategies, design methodologies, and voltage swing are the measures at circuit level. Parallelism, pipelining, and bus topologies are used as parameters at the architecture level. Data processing algorithms which minimize the switching activity for a given task are considered at the system level. High performance portable equipment, power-aware design techniques are desired to maximize the performance under power dissipation scenario. These methods can also help to reduce the energy cost. Hence power-aware verification plays an important role at present since semiconductor processing technology is continuously scaling down in terms of voltage and size with the aim of achieving performance, functionality at low power. In this work, various power management design techniques and the specification of power state tables (PSTs), switching networks, signal isolation, state retention, and restoration of APB protocol are introduced. APB is low performance and low bandwidth bus used to connect the peripherals like UART, keypad, timer, etc., to the bus architecture.

7.1   INTRODUCTION

Electronic applications are largely miniaturized, transformed from large computers to the system on chip (SoC)-based handheld portable devices, especially in medical, mobile, and entertainment applications. These devices also have long battery life with many features/functionalities available on a single SoC. These are possible due to advances in the semiconductor processing industry in terms of scaling feature size and supply voltage, which achieves exponential growth in chip density and more and more functions per unit area. Chip density, the number of transistors doubles every 18 months [16], i.e., a two-fold number of transistors are placed in the same die area impacts the overall power consumption increases by 2.7 times [2]. Technology scaling affects the supply voltage, which will be decreased not at the pace of an increase in chip density [4]. The frequency of operation is also impacting the chip power consumption. The general purpose and DSP processors operating at GHz speed are prone to a large number of switching activities, short circuits, and leakage currents, leading to the static and dynamic power dissipation and generation of heat within the device. The rise in temperature causes malfunctioning of the semiconductor devices, which in turn affects the reliability. Temperature is controlled by incorporating fans, blowers, and cooling mechanisms but which increases the design cost. Real-world computation and mobile applications are operating at high performance and also contain large density circuits consuming a lot of power. Today's VLSI design space was converted as three-dimensional since power has taken a fore front compared to performance and area. Hence, there are a lot of scopes to propose the various research solutions for power consumption minimization in the case of high speed and high-density chips. This work focuses on the proposal of low power CMOS design and standard power verification methodologies for complex SoCs, which are operating at high frequencies and with high density where power consumption can be within acceptable limits.

7.2 MATHEMATICAL MODELING

CMOS technology is preferred for the implementation of many applications due to low power consumption in stable logic states. Static power dissipation is zero, since at any time either p or n block only conduct. The major contribution is dynamic which will occur during the transition from logic 0 to logic 1 and vice versa. But in the case of high-density chips which are operating at reduced supply, channel length and threshold voltages Static power dissipation was observed in stable states due to leakage currents. Hence the total power dissipation of a CMOS inverter is the amalgam of static and dynamic components, can be quantitatively represented as in Eqn. (7.1):

$$P_{Total} = P_{Static} + P_{Dynamic} = P_{Leakage} + P_{Short\_Circuit} + P_{Switching} \qquad (7.1)$$

7.2.1 DYNAMIC POWER DISSIPATION

In CMOS devices dynamic power component is due to the load and transistors power consumption occurs because of transitions at input and output nodes and also due simultaneous on and off P and N blocks circuitry as shown in Figure 7.1.

7.2.2 SIGNAL SWITCHING POWER

Whenever input signal switches from logic 1 to logic 0, p device is on, current flowing from Vdd to C1, charges load capacitor Vdd changes the output from logic 0 to logic 1 results in power rise at the output as shown in Figure 7.1(a). In a similar manner if an input switch from logic 0 to logic 1, NMOS device is on, C1, discharges from logic 1 to logic 0 results to power fall at the output. Switching power is the energy required to charge/discharge the capacitor and independent of the duration of fall and rise times.

Switching power of a CMOS gate is quantitatively represented in Eqn. (7.2):

$$P_{switching} = \alpha \times C_1 \times V_{dd}^2 \times f \tag{7.2}$$

where; f is the switching frequency; α is the node transition parameter, effective number of power-consuming voltage transitions experienced by node per clock cycle and C_1 is the load capacitance.

7.2.3 SHORT CIRCUIT POWER

Short circuit currents are due to simultaneous on and off of 'p' and 'n' blocks during transition as shown in Figure 7.1(b). It is a considerable parameter when rise and false times are large and load capacitance is small.

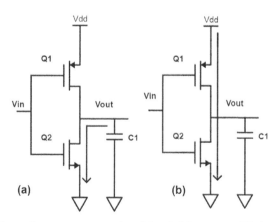

FIGURE 7.1   Dynamic power consumption: (a) switching power; (b) short circuit power.

7.3   STATIC POWER

In the case of submicron CMOS inverter, the NMOS and PMOS transistors are having nonzero reverse leakage and subthreshold currents. In a high-density CMOS VLSI chip transistors are operating near sub-threshold voltages considerable power dissipation is observed due to the leakage currents and which is determined by the processing parameters. System nodes which are under 100 nm contribute leakage power and it is around 40% in case of 65 nm node. In case of short channel devices, the channel length is the same order of magnitude as the depletion-layer widths of the source and drain junction [7]. Many short channel leakages such as drain-induced punch-through currents, surface scattering, hot carriers' injection from gate, and inverse leakage currents from drain and substrate, etc., will be observed as shown in Figure 7.2. To enhance the battery life of appliances, measures are required to reduce the leakage power.

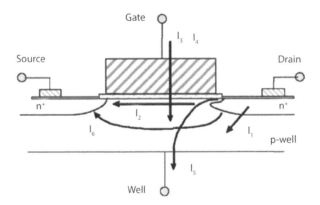

FIGURE 7.2   Transistor leakage currents.

The reverse leakage current is determined quantitatively using Eqn. (7.3):

$$I_{lk}a = I_s(e^{qVdd/kT} - 1) \qquad (7.3)$$

7.4   LOW POWER DESIGN

Currently, many real time electronic appliances like mobiles, wireless network interfaces require low power consumption to sustain for a longer time. There are many techniques developed during the last 10 years to satisfy the power requirements of SoCs and ASICs [5]. Performance of chip also affects the power consumption. Enhanced speed of operation of electronic devices increases the power consumption of the chip, since dynamic power is directly proportional to the frequency of operation. Though static power consumption is zero for CMOS devices, due to scaling there exist reverse leakage currents of drain and gate and short cannel, subthreshold and hot carrier injection (HCI) currents, etc., plays a vital in high density chip. Leakage currents can be reduced by reducing the supply voltage for proper operation. The major component of power dissipation in CMOS is dynamic power which is due to device and load conduction and signal switching. This dynamic power can be reduced by reducing the switching transitions and adjusting the operational frequency. The above stated issues lead to the proper design of CMOS low power and low voltage methodologies. SoCs with these features enhances the complexity. Moreover, life of the battery depends upon the power drawn by the system. Hence power consumption of chip should be kept within the acceptable limits to enhance the life of the devices. In this section few important methods such as substrate biasing, "clock gating, multi-switching (multi-Vt) threshold transistors, dynamic voltage and frequency scaling (DVFS), and unified power format (UPF)" [2], voltage scaling, up to sub one volt and RTL-based techniques are explained to reduce static and dynamic power.

7.4.1   CLOCK GATING

In this method, a gated clock, as shown in Figure 7.3, is applied, which provides a way to stop the original circuit to make transition at the subsequent redundant clock cycle. This method minimizes the dynamic power. In the case of reactive circuits, the idle state exists for a longer range of clock cycles [6].

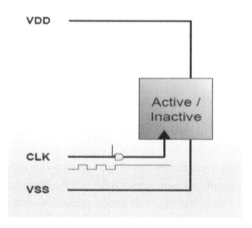

FIGURE 7.3   Clock gating.

7.4.2   POWER GATING

Power gating is also known as power down technique, reduces the power consumption by putting off the contemporary unused portion of the circuit as shown in Figure 7.4. "Shutting down the blocks can be accomplished either by software program or hardware" [2]. Hardware timers are also used. It affects the design and architecture of the circuit more than clock gating. Timing delays are also more in this method since power gated modes are to be correctly entered and executed [6].

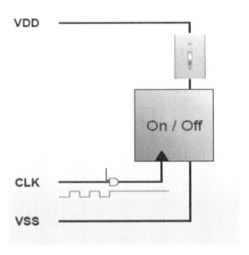

FIGURE 7.4   Power gating.

7.4.3   MULTI-VOLTAGE DESIGN

Nowadays in many chips, core, and output circuitry are operating at different supply voltages. These chips use multi-threshold CMOS (MTCMOS) transistors (V_{th}) to implement the assigned functionality with optimal power is shown in Figure 7.5. Low V_{th} devices with low swing switch faster are beneficial to be placed in the critical path to minimize clock delays. "But the disadvantage with the low V_{th} devices is prone to higher static leakage power. Hence high V_{th} devices are used in non-critical paths to reduce static leakage power without having much delay" [2]. Typical large V_{th} devices have 10 times less static leakage compare to with low V_{th} devices.

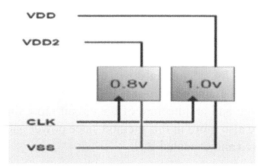

FIGURE 7.5 Multi-voltage design.

7.4.4 VOLTAGE/FREQUENCY SCALING

Dynamic power is proportional to the frequency. Frequency scaling shown in Figure 7.6 is used to adjust the frequency of a microprocessor dynamically either to conserve power or to reduce the heat generated by the chip. This method is also known as CPU throttling used in laptops and mobiles [6].

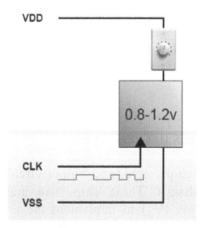

FIGURE 7.6 Voltage/frequency scaling.

7.5 NEED OF VERIFICATION

Semiconductor market demands for sophisticated ICs and in small packages. The main objective of functional verification is early detection of faults and to determine the solutions which are mapped into the ICs. The role of verification engineers is critical compared to RTL designers. The verification methods are discussed in further sections.

7.6   UPF OR IEEE 1801

UPF is proposed to meet the power objectives intended for electronic systems and designs. Since servers in information centers are used for applications such as the internet of things (IoT), electronic systems make use of enormous amount of energy on each day. Managing power demand through power gating and additional related techniques is needed in various electronic systems, specially to minimize heat and cooling cost as well as enhancing battery life. Power management involves partitioning a design into autonomous power domains and possible to operate in different power modes that optimize power utilization for every functional mode of the design. The UPF standard recommends a portable, vendor independent format essential for the management of power in structural design of a system to perform verification as well as accomplishment of the design flow aspects. The UPF is an IEEE 1801 standard and developed by Accellera. The prime objective of UPF is to design, simulate, and verify the chip designs having more power states and islands very easily. It is nothing but a power architecture which consists of power network and low power techniques. UPF enables power verification at the early stages of the design cycle, but it should be consistent at any stage of the design cycle. designs also modified correspondingly at the respective stages.

The UPF plays a key role in case of dynamic and static power issues in the CMOS low power process technology. A higher process node is more important since many functions are incorporated in a minimal area at low price. But it creates a power leakage since it needs minimum gate-to-source voltage differential to have conduction between drain and source terminals are creates a limit. Device supply voltage and its switching frequency contribute to dynamic power whereas supply, geometries, doping, and leakage currents directly contribute to a leakage power.

The emergence of new generation technologies with small threshold voltage devices causes the leakage power with increased static power dissipation [3]. To realize high performance chips, lower threshold voltage devices are preferred, but they are responsible to the increase of leakage power. At lower technologies like 65 nm [5] leakage power is a major concern in addition to the dynamic power. Hence supply voltage scaling is an important power management technique used to control the leakage and dynamic power components. Hence at present people are putting efforts on developing methods with voltage as a design parameter. Power reduction techniques such as multi-voltage, power gating, clock gating, dynamic voltage, and frequency scaling and management techniques are based on scaling and also power distribution network across the chip. In case of SoCs, power reduction methods are not sufficient, hence power verification methods are also necessary can be introduced at any abstraction level such as transistor, gate, or RTL.

7.7 UPF: FUNDAMENTAL MODELING CONSTRUCTS

UPF is the power management and verification methodology facilitate the incorporation of various power reduction techniques required for the formulation, modeling, and mapping of the power specification into the design. General UPF specification covers requirements such as the number and names of power domains and their constituent elements in terms of HDL instances, power distribution network type for the proposed design, corresponding states and customized specifications based on the considerations and conditions of the design. VLSI power design flow is shown in Figure 7.7. From Figure 7.7, it was observed that power files are also integrated with the HDL source files in the project. Both RTL source and power files play an important role to describe the design. Design RTL and UPF files are the input to all the tools of the flow such as simulator, synthesizer, place, and route, etc. Synthesis tool reads both the RTL and UPF design input files and produce a netlist. New UPF file can be generated by combining with the netlist, automatically or manually. PDF file which was inputted can be reused without changes at later stages of the design flow or alternate UPF files may also be used. In such case name of a UPF file need to be changed. Like DRC, UPF-aware logical equivalence checker can read the full design file perform the checks including the results of the UPF commands to ensure equivalence. Place and Route tool read both the netlist and the UPF files and produces the layout [10].

Power aware designs are introduced and verified at the early stages of the design cycle. In old designs power verification was implemented at the physical design stage since those designs have less no of pins and connections. It was known that all the design was available in on condition for all the time since power consumption of the chip was within the limits. Hence power supply network was simple with supply ports and nets but no other complicated power reduction techniques are not placed in the power designs. At present due to high density, low voltage and high-speed compactness of ICS, power-aware methods are to be included along with the power pins at the earlier stages due to complexity of layout. The power methods require additional circuitry and separate cells like isolation cells, level shifter cells, power switches, etc., in the design as well as during placement. At any point of time, all the blocks are not active; hence they are to be kept in power-down mode by applying gating techniques. Isolation blocks are inserted between blocks for valid transmission. In case of denser ICS core and peripheral circuits are operated at different voltages. To maintain the proper signal transmission strength level shifters are used between them. The power reduction methods can not disturb the functionality since they are switching between different modes and also at various voltage and frequency levels. Power design was introduced at the later stages fault detection is also a big problem hence designers are introducing power verification at the RTL level itself.

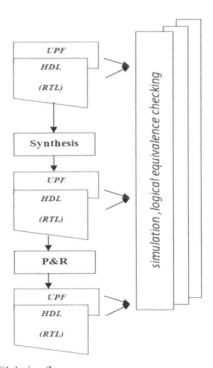

FIGURE 7.7   UPF VLSI design flow.

In conventional UPF design flow, there were few practical challenges such as errors are not identified at the early stage hence might lead to the wrong implementation of the power design. In addition to that, the UPF standard has a limitation of parallel development of power circuitry for complex designs. PDF file has to capture the Power-Aware characteristics of the design at the RTL/gate level in a compact form by the simulator. The scope of the UPF file is to make available in standard format to denote the supply network, switches, isolation, data retention, etc. UPF constructs are fundamentally classified into the following categories:

- Design scope;
- Primary power ports and primary ground;
- Power supply and supply networks;
- Power domains, interfaces, and boundaries;
- Power strategies;
- Power states and modes of operations.

These are the primary constituent parts used to model the power design like HDL constructs with the original design. The design methodology includes hierarchical top-down design module/instance names and power domains whose boundaries are defined by the hierarchical instance paths. Language reference manual (LRM) contains "the syntax and semantics of UPF constituent parts ensures accurate definition and with the inherent logical and lexical meaning of the defined constructs. UPF is the driving force for all the PA design verification and implementation automation tools. These tools are able to interpret and analyze the UPF fundamental constructs according to source and sink communication models which support inter or intradomain communication, strategy association, special power-aware cell such as ISO, LS, etc." [1]. insertions, deployment of corruption models, debugging, reporting of results, etc. HDL instances, ports, and nets are used to infer corresponding cells, supply, and control signals.

The implementation of fundamental components in UPF are described below in UPF 3.0. The backbone of UPF is TCL script only.

1. **Top Power Domain:** As in case of design the Top Power domain is created as follows:

   create_power_domainpd_SoC

2. **Primary Power Ports and Primary Ground:** Primary power ports and ground are implemented as follows:

   create_supply_portvdd_soc-domain pd_soc

create_supply_portvdd_soc_low-domain pd_soc
create_supply_portvss_soc-domain pd_soc

3. **Power Supply and Supply Networks:** Various supply nets for power and ground are specified below:
"create_supply_netvdd_soc_n-domain pd_soc
create_supply_netvdd_soc_low_n-domain pd_soc
create_supply_netvss_soc_n-domain pd_soc
#connect the supply ports to the supply nets
connect_supply_netvdd_soc_n-ports {vdd_soc}
connect_supply_netvdd_soc_low_n-ports {vdd_soc_low}
connect_supply_netvss_soc_n-ports {vss_soc}" [5].

4. **Power Domain:** This used for various primary supply net are mentioned below.
"set_domain_supply_netpd_soc \
-primary_power_netvdd_soc_n \
-primary_ground_netvss_soc_n
#create switchable power domains
create_power_domain PD_slave0-elements s0
#specify the supply ports for power and ground
create_supply_port Vslave0_port-domain PD_slave0
create_supply_port Gslave0_port-domain PD_slave0" [10].

5. **Isolation Cells:** These cells are placed between two blocks can be independently powered off. When a design block is powered off, the isolation cell clamps its output to a predetermined value so that the active block is protected from undesired signals and interrupts from the powered off block. Isolation blocks are created as follows.
"#isolation strategy for UART power domain
set_isolation iso_slave0 \
domain PD_slave0 \
isolation_power_net Vslave0_net \
clamp_value 0 \
applies_to outputs
#isolation control
set_isolation_control iso_slave0 \
-domain PD_slave0 \
-isolation_signal slave0_iso_en \
-isolation_sense high" [8]

6. **Level Shifters:** "They convert the voltage range of an output signal to a different voltage range suitable for the input logic of the receiving block. The level shifter is created as follows" [4].

#set the level shifter

set_level_shifterapb_shifter-domain pd_soc \

applies_to outputs-location self-rule both

7. **Retention Registers:** These are used to preserve the state history of a power domain during on and off transition. The memory cell which was not initialized/reset will have an unpredictable initial value represented 'X.' UPF code to create retention register is mentioned below.

"retention specification for UART power domain

set_retention slave0_ret \

-domain PD_slave0 \

-elements {s0} \

-retention_power_net Vslave0_net

#retention control specification

set_retention_control slave0_ret \

-domain PD_slave0 \

-save signal {ret_PD_slave0 posedge} \

-restore_signal {ret_PD_slave0 negedge}" [2].

8. **Power Switches:** It controls the power flow from supply to power domains. They are managed by the signals sourced from a power management unit within the SoC. Whenever they receive a request, they establish a flow of current to the concern power domain. The following code explains creation of a power switch for a slave power domain.

"create_power_switch slave0_sw-domain PD_slave0 \

-output_supply_port {vout_p slave0_primary_power} \

-input_supply_port {vin_p Vslave0_net} \

-control_port {ctrl_p PD_slave0_cntl} \

-on_state {full_onvin_p {ctrl_p}}" [1].

9. **Power Supply Network:** "The concept of power domains and multi-voltage creates a need for a network and hierarchy of power lines, switches, and on-chip and off-chip power regulators. Since some of the power lines can be controlled to supply different voltage levels at different times, care must be taken when coordinating the voltage levels with clock frequencies for a given power state" [2].

#specify the supply nets for power and ground
"create_supply_net Vslave0_net-domain PD_slave0
create_supply_net Gslave0_net-domain PD_slave0
create_supply_net slave0_primary_power-domain PD_slave0
#connect the supply ports to the supply nets
connect_supply_net Vslave0_net-ports {Vslave0_port}
connect_supply_net Gslave0_net-ports {Gslave0_port}
#specify the domain primary supply net
set_domain_supply_net PD_slave0 \
-primary_power_net slave0_primary_power \
-primary_ground_net Vslave0_net" [5].

7.8   APB DESCRIPTION

This section describes the modeling and implementation of AMBA in UPF. it is a part of APB bus and also be the subshell of the AMBA architecture. It defines a low-cost interface with the minimal power consumption. AMBA design includes direct memory access (DMA) or digital signal processor (DSP) as bus masters, and the test circuitry. In general, external memory interface, the advanced peripheral bus (APB) bridge and internal memory, etc., as the common AHB slaves. "It was not pipelined and is used to interface the low-bandwidth peripherals which are not necessary for high performance AXI protocol. The APB protocol synchronizes signal transition to the rising edge of the clock, to simplify the integration of APB peripherals into any design. Each transfer needs at least two cycles" [6].

7.8.1   APB PROTOCOL

The APB is a component of the AMBA performs a sequence of significance transfers at low power consumption and minimize interface environment. The APB is used to interface several peripherals which are at low data transmission and without pipelined transfer interface. In APB all signals are synchronized with the rising edge of the clock [4, 5]. The advanced RISC machine (ARM) microprocessor is very popular for SoC solutions. Interfacing of two APB devices is shown in Figure 7.8.

7.9   APB PROTOCOL POWER MANAGEMENT ARCHITECTURE

APB SOC, in addition to the normal power networks, ports, and additional low power techniques are also required to operate properly. Hence active low power management techniques such as power gating, multiple voltage supplies, partitioning the design into separate functional areas which can be incorporated into the SoC architecture can be powered ON/OFF independently. Additional logic such as power switching, isolation, level shifting, and state retention are to be inserted into the design to perform special functions. These additional components contribute to the power management architecture for a given system.

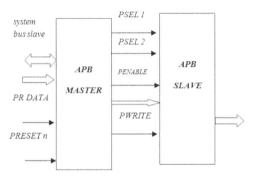

FIGURE 7.8   Interfacing of APB master and slave.

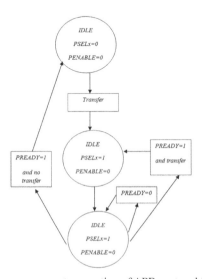

FIGURE 7.9   State diagram represents operation of APB protocol [3].

7.9.1   APB IMPLEMENTATION

Implementation of APB in UPF can be carried out using two modules known as master and slave. Either of them can be powered ON/OFF independently. Both master and slave outputs are protected from wrong signals from a powered OFF driver by the use of isolation cells. Isolation cells always gets a constant supply. "The top power management module captures the control signals in order to close the power domain switches. It also drives the retention signals to retain the memory and register content on whenever required" [5]. In this work, APB is implemented as a top module as shown in Figure 7.10, which contains a master and slave as submodules. Master executes read-write operation by a slave. During a write operation, the master writes to the particular slave addressed by the address line. During the read operation, the master enables the address of slave from which it wants to read the data.

7.9.2   OPERATING MODES

Power management design for a given system involves the functional description and operating modes of the system. The aim of the active power management system is to optimize the use of power based on functions and also performance of the APB system at any time. The first step is the recognition of distinct combinations and the no of combinations of functions are required for proper use. Partitioning of the SoC is performed based on the set of individual operating modes which independently powered on/off subsystems/subcomponents, so that any given operating mode can be supported by providing the necessary subsystem with the appropriate power. For example, block diagram of a design shown in Figure 7.10 has two operating modes ON and OFF. In the proposed architecture APB top module which contains a master and slave as subsystems. "UPF file is intended for the complete design considering both as switchable modules in order to preserve the power. The top architecture contains two power domains including top module. These are switchable according to conditions written in power state table (PST) using power switch domain. The top power domain works on 0.99 V and 0.81 V correspondingly. Retention cells are developed and shown in the Top Power module are useful to save values before particular domain gets switched OFF and restore the same after domain gets switched ON. The sequence of UPF commands in the power-aware design are represented in Figure 7.11 used to perform operations on the top module in order to conserve the power [1].

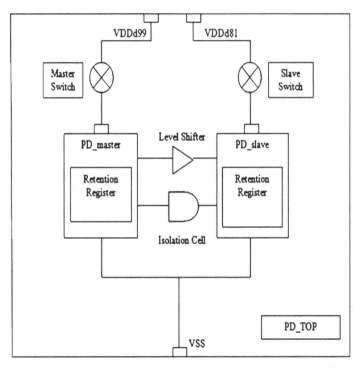

FIGURE 7.10   RTL implementation of APB top module architecture with UPF.

> ➤ set_design_top
> ➤ set scope
> ➤ create_power_domain
> ➤ creare_supply_port
> ➤ create_supply_net
> ➤ connect_supply_net–port
> ➤ create_power_switch
> ➤ set_isolation
> ➤ set_isolation_control
> ➤ set_retention
> ➤ set_retention_control
> ➤ set_level shifter

FIGURE 7.11   UPF commands operated on top power module file [7].

7.10   SIMULATION RESULTS

In this work APB protocol was implemented without power-aware and with Power-aware design (UPF) using HDL and compared. The simulation results are shown in Figures 7.12 and 7.13, respectively. All the states such as isolation, power modes are observed in Figure 7.13 with power aware states. This work concentrates on the implementation of APB protocol in the presence of power aware control signals and the UPF file in the design. While implementing power control signals such as switch ON and OFF for each power domain, isolation signal for the protection of each power domains when power on and off, retention signal to save the data before power off to the power domain and level shifter signal to convert the voltage high to low and vice versa between various parts of the circuit. Switch signal is used to switch between normal and power-aware modes whenever it is asserted to low, the assigned power domain is moved to the off state at the time save signal is enabled in order to save the relevant data present at that time. "During this period isolation signal is enabled to isolate the power domain from the other power domains, so that it doesn't affect by the other domains. When the switch signal is high, then assigned power domain will be in the on state, then the restore signal is enabled and saves the data, after few times periods isolation signal will be low to remove isolation" [1].

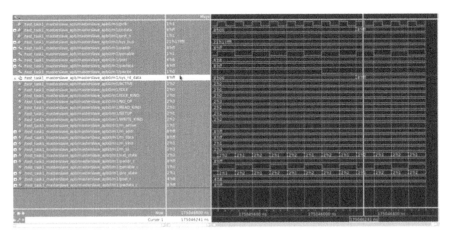

FIGURE 7.12   Simulation results of APB without UPF.

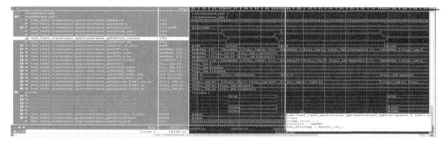

FIGURE 7.13   Simulation results of APB with UPF.

Finite state machine (FSM) shown in Figure 7.14 corresponds to the PST. It is a table that represents the permitted combinations of power states for those set of supply nets/supply ports.

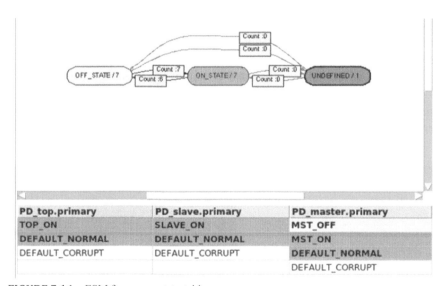

PD_top.primary	PD_slave.primary	PD_master.primary
TOP_ON	SLAVE_ON	MST_OFF
DEFAULT_NORMAL	DEFAULT_NORMAL	MST_ON
DEFAULT_CORRUPT	DEFAULT_CORRUPT	DEFAULT_NORMAL
		DEFAULT_CORRUPT

FIGURE 7.14   FSM for power state table.

7.11   CONCLUSION

Active power management is a fundamental problem that needs to be addressed, especially in case of high density, small size, and high performance SoCs are the integral part of many applications such as mobile, health, and network applications where power dissipation is high due to high switching rate. To integrate active power management methods into a complex SoC designs and ensuring that they were working correctly and also not affecting the original functionality of the chip is also an important task. The methodology needs to define, build, and verify the appropriate power management architecture. IEEE Standard 1801 UPF is PA methodology perform static analysis of power management architecture, power-aware simulation of power managed designs, and formal verification of power control logic. These methods provide a comprehensive solution for defining and verifying active power management.

ACKNOWLEDGMENT

The authors would like to convey their sincere gratitude to Mr. Srinivasan Venkataramanan of CVC Pvt. Ltd, Bangalore for the tools and technical support provided to carry out this work.

KEYWORDS

- **AMBA**
- **dynamic voltage and frequency scaling**
- **internet of things**
- **low power**
- **power aware simulation**
- **system on chip**
- **unified power format**

REFERENCES

Bembar, F., Kakkar, S., Mukherjee, R., & Srivastava, A., (2009). Low power verification methodology using UPF. In: *Conference on Electronic Systems Design and Verification Solutions* (pp. 228–233). DVCON.

Chadha, R., & Bhasker, J., (2013). *An ASIC Low Power Prime Analyzer* (1st edn.). Springer Company.

Croft, M., & Bailey, S., (2007). Is your low power design switched on? *International Symposium on System-on-Chip*, 1–4.

Design Automation Standards Committee of the IEEE Computer Society, (2015). *IEEE Standard for Design and Verification of Low-Power, Energy-Aware Electronic Systems.* IEEE Std 1801-2015.

Hazra, A., Mitra, S., Dasgupta, P., Pal, A., Bagchi, D., & Guha, K., (2010). Leveraging UPF-extracted assertions for modeling and formal verification of architectural power intent. In: *47th ACM/IEEE Design Automation Conference (DAC)* (pp. 773–776).

Hello, S., Shah, H., Modi, C., & Tarpara, B., & Chinmay, M., (2015). *Design and Implementation of Advanced Peripheral Bus Protocol (IJSEAS)* (Vol. 1, No. 3.). ISSN: 2395-3470.

Hyo-Sig, W., Kyo, S. K., Kwang-Ok, J., Ki-Tae, P., Kyu-Myung, C., & Jeong-Taek, K., (2003). An MTCMOS design methodology and its application to mobile computing. *Low Power Electronics and Design, 2003. ISLPED '03; Proceedings of the 2003 International Symposium*, 110–115.

IEEE 1801™- (2009). Standard for design and verification of low power integrated circuits. IEEE. Stephen Bailey, Gabriel Chidolue, Allan Crone, of mentor graphics. *Low Power Design and Verification Techniques*.

Jian, H., & Xubang, S., (2008). The design methodology and practice of low power SoC. In: *Embedded Software and Systems Symposia, 2008* (pp. 185–190). ICESS Symposia'08. International Conference.

Kapoor, B., Hemmady, S., Verma, S., Roy, K., & D'Abreu, M. A., (2009). Impact of SoC power management techniques on verification and testing. *Quality of Electronic Design* (pp. 692–695). ASKED.

Khondkar, P., & Bhargava, M., (2016). *The Fundamental Power States: The Core of UPF Modeling and Power-Aware Verification*. Whitepaper at Mentor.com (accessed on 21 September 2021).

Khondkar, P., (2016). Power-aware libraries: Standardization and requirements for Questa® power-aware. *Verification Horizons Journal, 12*(3), 28–34.

Khondkar, P., (2017). *The Concepts and Fundamentals of Power-Aware Verification* (1st edn.). New York, Springer, communicated.

Khondkar, P., Yeung, P., et al., (2017). *Free Yourself from the Tyranny of Power State Tables with Incrementally Refinable UPF.* DVCon.

Khondkar, P., Yeung, P., Prasad, D., Chidolue, G., & Bhargava, M., (2016). Crafting power-aware coverage: Verification closure with UPF IEEE 1801. *Journal of VLSI Design and Verification, 1*, 6–17.

Kim, N. S., Austin, T., Blaauw, T., Mudge, T., Flautner, K., Hu, H. S., Irwin, M. J., et al., (2003). Leakage current: Moore's law meets static power. *IEEE Computer, 36*(12), 68–75.

Krishnamurthy, Atila, A., Vivek, D., & Shekhar, B., (2002). High-performance and low-power challenges for sub-70 nm microprocessor circuits. CICC. *IEEE Proceedings*, 125–128.

Lissel, R., & Gerlach, J., (2007). Introducing new verification methods into a company's design flow: An industrial user's point of view. In: *Design, Automation and Test in Europe, Conference & Exhibition DATE'07* (pp. 689–694). IEEE.

Mehta, S., (2008). *Industry Standards from Accellera 21st International Conference on VLSI Design, VLSID*, 728.

Neil, W., & David, H., (2005). *CMOS VLSI Design: A Circuits and Systems Perspective* (4th edn.). Addison-Wesley Publishing Company.

Rudra, M., Amit, S., & Stephen, B., (n.d.). *Static and Formal Verification of Power-Aware Designs at the RTL Using UPF.*

Shigematsu, S., Mutoh, S., Matsuya, Y., Tanabe, Y., & Yamada, J., (1997). A 1-V high-speed MTCMOS circuit scheme for power- down application circuits. *IEEE J. Solid-State Circuits, 32*(6), 861–869.

Trummer, C., Kirchsteiger, C. M., Steger, C., Weiss, R., & Dalton, D., (2009). *Simulation-based Verification of Power Aware System-on-Chip Designs Using UPF IEEE 1801* (pp. 1–4). NORCHP.

Trummer, C., Kirchsteiger, C. M., Steger, C., Weiß, R., Pistauer, & Dalton, M. D., (2010). Automated simulation-based verification of power requirements for systems-on-chips. In: *IEEE 13*[th] *International Symposium on Design and Diagnostics of Electronic Circuits and Systems (DDECS)* (pp. 8–11_.

Zyuban, V., & Kosonocky, S. V., (2002). Low power integrated scan-retention mechanism. Low power electronics and design, 2002. ISLPED '02. *Proceedings of the 2000 International Symposium*, 98–102.

CHAPTER 8

Architecture Design: Network-on-Chip

N. ASHOK KUMAR,[1] A. KAVITHA,[2] P. VENKATRAMANA,[3] and
DURGESH NANDAN[4]

[1]*Department of Electronics and Communication Engineering,
Sree Vidyanikethan Engineering College, Tirupati, Andhra Pradesh, India,
E-mail: ashoknoc@gmail.com*

[2]*Veltech Multitech Dr. Rangarajan and Dr. Sankuntala Engineering
College, Chennai, Tamil Nadu, India*

[3]*Sree Vidyanikethan Engineering College, Tirupati, Andhra Pradesh, India*

[4]*Aditya Engineering College, Surampalem, Andhra Pradesh, India*

ABSTRACT

Network-on-chip (NoC) structure indicates a capable pattern concept to manage with growing data transfer needs in digital process. NoC has been come out as a strong aspect that decides the functioning and power utilization of several principal processes. VLSI technology is utilized to change NOC internal router arrangements, shortest route assigning methods and neighbor router assessment control. This minimizes the route assigning process time and to improve NOC structure functioning stage. The present system creates a mesh topology-based network on chip structure. This structure executes the circuit exchanging and blocks of data exchanging for route assigning process. Present blocks of data exchanging method are to detect the router accessibility for mesh topology-based NOC structure pattern and circuit exchanging method is utilized to detect the route presence for mesh topology NoC structure. The present system optimizes the route assigning time and effectively transmits the data from the source to destination. Present system increases the circuit complexity stage and it consumes more time for circuit analysis process. Proposed system creates a mesh topology depend router structure pattern and minimize the route sharing methods using hybrid system.

8.1   BLOCK OF DATA FORMAT

The network-on-chip (NoC) consists of objects which are related through links with a data width of the sequence of data that is transported across a network [1]. Each transmission of a data element on such a hyperlink is referred as NoC sequence of data that is transported across a network. To permit the switch of big sized units of memories a couple of sequence of data that is transported across a network can be merged to NoC block of data. Hence, a sequence of data that is transported across a is drawn out with two more bits for the type (FLIT_TYPE_WIDTH) which are brought in at the MSB [1]. The bits denotes whether this is the first sequence of data that is transported across a network in a NoC block of data (bit 0) and/or this is the last sequence of data that is transported across a network in a NoC block of data (bit 1). As we use wormhole routing, such block of data is the indivisible routing block. If you have a 32-bit data width that means, that a sequence of data that is transported across a network is 34-bit wide. Those sequences of data that is transported across a network create a block of data by way of setting the type bits correspondingly. A block of data of length N has the sequence of data that is transported across a network <01, sequence of data that is transported across a network 0>, N–2 times <00, sequence of data that is transported across a network i>, and finally <10, sequence of data that is transported across a network N–1>. It is additionally feasible to have a block of data of size 1 (<11, sequence of data that is transported across a network 0>).

The first sequence of data that is transported across a network steadily has the limited location of the block of data header. The NoC itself solely needs that the first PH_DEST_WIDTH comprises the target id. This is utilized to do the routing search for (as we do dispense routing). The other bits of the header can be utilized freely, and the associated blocks of course use those bits (Figure 8.1).

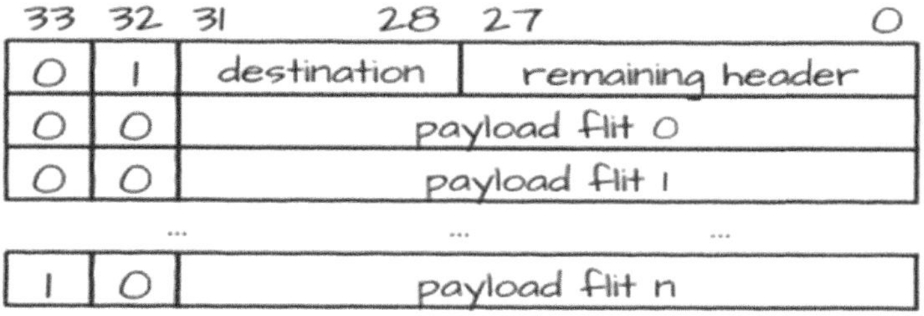

FIGURE 8.1   Block of data format.

Summarized the NoC has two factors that outline the block of data organization:

SEQUENCE OF DATA THAT IS TRANSPORTED ACROSS A NETWORK_DATA_WIDTH-

Width of the true sequence of data that is transported across a network data.

PH_DEST_WIDTH-

Quantity of most giant bits in the header that decide the break spot of a block of data. For SEQUENCE OF DATA THAT IS TRANSPORTED ACROSS A NETWORK_DATA_WIDTH=32 and a NoC with up to 16 associated blocks (PH_DEST_WIDTH=4) the NoC block of data are created like this.

8.2   EXCHANGING METHODS

Once the arrangement of the elements of the NoC is decided, the exchanging method, or how statistics moves in the data forwarding devices, should be found out. This concerned describing the granularity of facts switch and the exchanging method [11]. Data is transmitted on a connection, which has a constant width, recorded in bits. The quantity of records transmitted in one cycle on a connection is known to as the physical units (physical unit). Two data forwarding devices coordinates every data transmission, to assure that buffers do no longer spread out. Connection-level waft control is utilized for this, and can be depends totally on handshaking or the use of credit [5]. A part of the synchronization is known as a sequence of data that is transported across a network (flow manipulate unit), and it is at least as large as a physical unit. At last, a couple of sequence of data that is transported across a network comprise a block of data, many of which may additionally create messages that blocks associated to the NoC and communicate between them [12]. Messages can be utilized for exceptional NoC programming concepts, along with message transferring and distributed-shared memory (Figure 8.2 displays this configuration). To extend the block of dataization effectiveness message and block of data borders do not want to be organized, as shown in Figure 8.2.

Different NoCs use exclusive physical unit, sequence of data that is transported across a network, block of data, and message sizes. The physical unit and sequence of data that is transported across a network size reproduce various diagram selections, such as hyperlink speed versus muter arbitration pace [12]. For example, /Ethereal makes use of physical units of 32 bits, a sequence of data that is transported across a network of three physical units (or words), and a block of data and messages of unbounded length. Nostrum two utilizes physical units of 128 bits, and sequence of data that is transported across a network equal to 1 physical unit. SPIN two utilizes physical units and sequence of data that is transported across a network of 36 bits, and block of data can be unbounded in length [5]. Physical units are applicable for the hyperlink layer, and will not be further described. The exchange method decides how the sequence of data that is transported across a network and block of data are transported and saved by way of the data forwarding devices. Note that the exchanging method decides how statistics flows via the NoC, moreover no longer alongside which path.

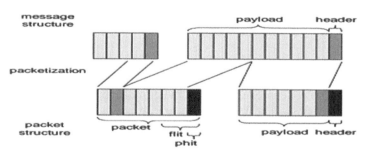

FIGURE 8.2   Physical units, sequence of data that is transported across a network, block of data and message.

Two major types of moving sequence of data that is transported across a network exists: design exchanging and block of data exchanging. Fundamentally, in design, exchanging a design with a constant physical path is created among transmitter and receiver, and all the sequence of data that is transported across a network of the message are sent out on this design. Alternatively, in block of data exchanging the block of data comprising a message creates their path and does not depend from transmitter to receiver, possibly along limited paths, and with surprising delays. We talk about the methods, and countless transforms, in a detailed manner as follows. To decide an exchanging method for an NoC, a diversity of problems have to be managed, as like the granularity of the facts to be transmitted, and the transmitting frequency; the cost and difficulty of the data forwarding device; the vitality and amount of simultaneous moves to be maintained; and the resultant actions (bandwidth, delay) of the NoC [12]. Several categories of

NoC frequently use one-of-a-kind exchanging methods. The exchanging method robustly controls QoS, in fact, to provide one-of-a-kind QoS layers, NoCs can use a couple of exchanging methods at the equal time.

8.2.1   CIRCUIT EXCHANGING

Data from one block to every other are transmitted in their entity when the usage of design exchanging. At the beginning, a bodily route, that is, a sequence of hyperlinks and data forwarding devices, from transmitting to receiving block is decided and allocated for the design. Reasonably, the head sequence of data that is transported across a network [1] of the message creates its path from the transmitter to receiver, preserve links beside the path. If it reaches at the receiver barring disagreements (all hyperlinks were existing), an acceptance is transmitted again to the transmitter, who begins data switch [11] on its reception. If a hyperlink is hold by way of any other design, a poor acceptance is sent to the transmitter. Figure 8.3 suggests the creation and utilizing a design over time; RI, R2, and R3 are data forwarding devices alongside the direction of the design. The sheltered boxes display if the inter-data forwarding device wires are engaged [12].

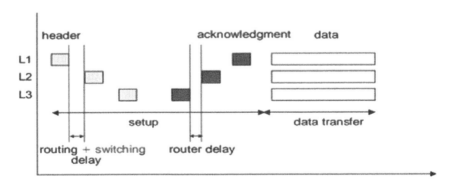

FIGURE 8.3   Circuit exchanging.

After the message transfer, the design is worn out, as a section of the tail sequence of data that is transported across a network. Pure design exchanging has no longer been utilized a whole lot in NoCs, and the main examples are SOC Bus. Circuit exchanging has an extreme primary delay because of the created area that should be finished before data transfer begins [9]. Scouting routing can decrease this time. Data transfer is more effective, conversely, the entire hyperlink bandwidth is on hand to the design, and consequences in reduced delay. Data should not be buffered in the data forwarding devices (only pipelined maybe), lowering the position of data forwarding devices [12]. Conversely, design exchanging does not ascend nicely as the size of the NoC develops as hyperlinks are engaged additionally if statistics is now not being transmitted at few points of the creation and worn-out stages.

Circuit exchanging is remarkable when data is sent out frequently, or if the verbal exchange style among transmitters and receivers is comparatively stationary. The design can be left in a neighborhood in these situations [5]. If the quantity of statistics to be transferred is huge (creating the creation segment not significant) design exchanging also performs good. ASICs and ASSPs have fairly constant conversation styles, and FPGAs additionally transmits data each cycle on the design. FPGAs completely use design exchanging and ASIC and ASSP NoCs frequently do.

8.2.1.1   FUNDAMENTAL PATHS AND FUNDAMENTAL CIRCUITS

Circuit exchanging preserves bodily connections among data forwarding devices. Several digital connections (more in many instances known to as digital paths) can be multiplexed on a no longer be sufficient Fundamental paths to hold all movements. Fundamental-design buffered data forwarding device implementations turn out to be complicated either because of multiplexing the huge category of small buffers, or because of the largely shared buffer executions, as like SRAMs. Data forwarding devices that put in force digital circuits [9] with a single buffer per hyperlink are unnecessary to say do no longer go through from this issue, as explained as follows. The multiplexing of the single Fundamental paths on a single hyperlink needs scheduling at all connection/data forwarding device, and produces an end-to-end schedule of the fundamental circuits [9]. Fundamental circuits with input queuing are utilized. The scheduling of get entry to connections and contact to the (stopping) crossbar in the data forwarding device interfere, creating bandwidth and delay assurance tough to obtain. Hence, the Mango NoC makes use of fundamental-channel buffering executed with the assist of output queuing with a non-blocking crossbar two (Figure 8.3). This, together with a fine hyperlink scheduling method builds it to assure bandwidth and delay on its digital circuits.

Fundamental circuits with one buffer per connection circuits can also be time-multiplexed with one buffer per connection, that is, excluding enough buffering per digital design. Basically, this is completed by means of statically scheduling (using TDM) the usage of all links in the NoC through all Fundamental circuits. Figure 8.4 proposes a 4 × 4 data forwarding device with four TDM slots [10]. Each enter has a one-sequence of data that is transported across a network buffer even though it can be utilized through up to 4 digital circuits. The TDM tables displays how digital circuits are mapped from inputs to output in time. At the part of the NoC, sequence of data that is transported across a network [1] are introduced by the network interfaces such that they by no means utilize the similar hyperlink at the equal time. Hence, connection-level go with the flow control and scheduling can be avoided, and solely one sequence of data that is transported across a network buffer per connection is mandatory. This method guesses the propagation velocity of man or woman sequence of data that is transported across a network via the NoC is steady and identified in advance. Sequence of data that is transported across a network wait in the network interfaces till they are injected in the NoC as per the TDM schedule. Note that community interfaces want some kind of end-to-end go with the flow control for lossless working, possible with fundamental-design buffering. The TDM is utilized to globally time table all flows, and as a result, end-to-end bandwidth and delay assurances are easily given [10]. The thinking of globally scheduling all fundamental circuits is used through NuMesh (for parallel processing), Nostrum's looped containers, adaptive SoCs (a SoCs) world scheduling, and Ethereal's contention-free routing.

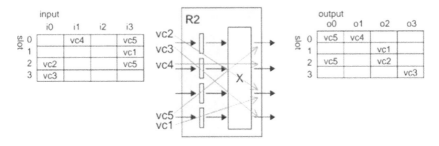

- input i0 transports two virtual circuits (vc2, vc3)
- input i1 transports one virtual circuit (vc4)
- input i3 transports two virtual circuits (v1, vc5)

FIGURE 8.4   Fundamental circuit exchanging with a single buffer per connection and TDM.

8.2.1.2   CIRCUIT MANAGEMENT

Creation and worn-out stages of digital circuits can consume area as explained at the starting of the subdivision. Options surround static or dynamic reservations, back-tracking, and multicast. The task should be monitored such that impasse does not happen now throughout set up. This can be done through retracting a design, or by losing statistics. In additionally, a programming can be centralized in one entity such as a programmable processor. Creation, (negative) acknowledgment, and tear-down can also be encoded as messages like in the ATM and pipelined design exchanging. Creation and worn-out message's application data forwarding devices, the use of a message-based or memory-mapped protocol. Ethereal and Mango are examples of this approach. Control messages can use a one of a kind exchanging schemes (e.g., wormhole (WI-I) block of data exchanging) and unique QoS [6] type. Using messages approves design management to use the NoC itself, casting off a separate control interconnect to configure the NoC.

8.2.2   BLOCK OF DATA EXCHANGING

In (fundamental) design exchanging a whole (distributed) direction is held before the data transmission. In contrast, in block of data exchanging, no hyperlink allocations are made, and the block of data comprising a message create their way separately from transmitter to receiver, possibly along distinctive paths, and with unique delays. Ignoring the creation phase eliminates the start-up time (creation until approval), however besides hyperlink reservations, block of data of one-of-a-kind flows may attempt to use a hyperlink at the equal time. This is known as contention, and requires that all however one block of data should wait till the hyperlink turns into accessible again [5]. The start-up waiting time of design exchanging (followed by way of a fixed minimal delay in the data forwarding devices) is changed by a zero-start-up time and a variable prolong due to contention in every data forwarding device alongside the block of data's path. Hence, QoS [6] ensures are more difficult to supply in block of data-switched NoCs than in design-switched NoCs [11]. Buffering needs of SAF exchanging and VCT exchanging are similar, and no NoC has utilized this primary method (Figures 8.5 and 8.6).

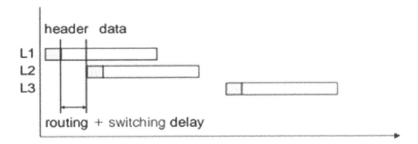

FIGURE 8.5   VCT exchanging.

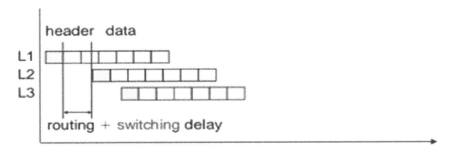

FIGURE 8.6   Wormhole (WH) exchanging.

8.2.2.1   WH EXCHANGING

WH exchanging enhances on VCT exchanging through minimizing the buffer necessities to one sequence of data that is transported across a network. This is accomplished by transferring each sequence of data that is transported across a network of a block of data when there is an area for the sequence of data that is transported across a network in the receiving muter (Figure 8.6). If no house is handy in the subsequent data forwarding device for the complete block of data (as required by way of VCT exchanging however no longer with the aid of WH exchanging), the block of data is connected over two or many data forwarding devices. This prevents the connection and hence results in higher congestion than with SAF and VCT exchanging. Connection obstruction can be improved by way of multiplexing fundamental links (or fundamental paths) on one bodily connection. WH exchanging is additionally greater inclined to impasse than SAF and VCT exchanging because of the recently initiated usage dependency among links. Fundamental paths and/or routing methods can be utilized to keep away from deadlock. Several NoCs use WH exchanging except digital paths utilize constrained topologies to keep away from deadlock. SPIN makes use of a fats tree arrangement with deflection routing and block of data reordering at the receiving community interface, which additionally get out of routing deadlock. Athereal's BE provider type utilizes WH exchanging class and prevents impasse for any arrangement by means of an aggregate of turn-prohibition routing and end-to-end waft manipulate. Other processes, utilize WH exchanging with fundamental paths.

8.2.3   COMBINATIONS OF DIFFERENT EXCHANGING

Methods Circuit exchanging and block of data exchanging have distinct traits that may additionally be beneficial to mix in a single NoC. With design exchanging it is rather convenient to assurance bandwidth and delay assures for a given set of (fundamental) circuits. Block of data exchanging, on the different hand, has no idea of (fundamental) circuits and can consequently assist any number of concurrent flows. However; it is challenging to give difficult bandwidth and assured delay. Fundamental design exchanging and block of data exchanging can be blended through assigning Fundamental channel to each class. Assured services are ported to digital circuits and BE services to block of data. Ethereal, Nostrum, and Mango are employed this method. The first two use TDM and Mango makes use of "asynchronous delay assure" scheduling [10]. Note that though WIT exchanging with infinite block of data lengths also makes Fundamental circuits, however to offer end-to-end execution ensures the proper scheduling self-discipline is mandatory. Primary block of data exchanging is gorgeous for very dynamic applications where data transfers exist only for a short duration, alternate frequently, or transforms often in their commands, due to the absence of waft reservations, and creation and tear-down stages. Comparatively short messages (synchronization traffic, cache lines, etc.), to several distinct blocks (e.g., distributed reminiscences and caches) creates relationship-based QoS unacceptable [6]. Relationship-less block of data exchanging with BE service type is hence natural for CMP NoCs. Conversely, to provide confident service classes, relationship-based block of data exchanging, primarily based on (TDM) [10] digital circuits are used effectively for ASICs and ASSPs [8].

8.3 ASYNCHRONOUS FIFO IMPLEMENTATION

Distributing numerous bits, whether information bits or manage bits, can be completed by means of an asynchronous FIFO. An asynchronous FIFO is a distributed memories or register buffer in which, statistics is introduced from the write clock area and statistics is separated from the examiner clock area [16]. As both transmitter and receiver function inside their very own corresponding clock areas, utilizing a dual-port buffer, as like FIFO, is a secure method to transmission multi-bit values among clock domains. A general asynchronous FIFO gadget lets in numerous information or manipulate words to be introduced if the FIFO is free. The receiver can then remove out several information or manage words when the FIFO is no longer clear.

8.3.1 DESIGN OF AN ASYNCHRONOUS FIFO THROUGH GRAY CODE COUNTERS

The Gray Code counters are utilized in this asynchronous FIFO arrangement for the Read_ pointer and the Write_pointer assuring victorious switch of many-bit facts from write clock (also known as the transmit clock) to study clock (aka the acquire clock). Let us develop at an asynchronous FIFO format that uses Gray Code counter. For the design of the synchronizer, there will be a hypothesis on TxData steadiness. It should be safe for two to three clocks. This builds positive that the CDC sign RxI Data has sufficient time to filter the metastability vicinity and avoid the exact price to RxData. It is assumed that the TxData have to remain to be constant for two clocks each and every time it considers a new cost (i.e., it modifies). This guess is mandatory when you think about that we count on that TxClk is quicker than RxClk. The timing layout of this design is shown in Figure 8.7. Here is a minimal assertion to verify for TxData steadiness [7].

First, the announcement verifies that TxData has transformed at posedge of TxClk. TxData is maintained into the numerous threaded close by changeable data. I 'b I is necessary because of the fact neighborhood information collect have to be linked to a representation. As we do not have any situation, we essentially say "always true" is the representation. "Always true" potential continually saves TxData into the data, every time TxData transforms. Then, we take a look at the CDC boundary clock RxClk that the records have certainly transferred to Rx I Data through evaluating Rx IData with the saved TxData (in the data). One clock later, the RxData have to in shape the TxData that was once transferred on TxClk. This assures that the CDC I-bit two-flop synchronization performs as planned. Again, be aware that the guess of TxClk quicker than RxClk should be held on to. It is easy to declare to test for glitch on TxData. The above answer believes that TxData does not has glitches.

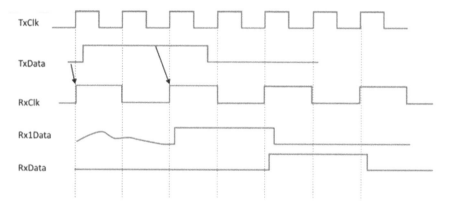

FIGURE 8.7   Lengthened transmit pulse for exact capture in receiver clock domain.

Let us consider a widespread declaration for a numerous-bit Gray Code counter-dependent statistics transmit diagonally CDC area. This declaration is recorded for the asynchronous FIFO layout. The write data are recorded to fifo_in on welk (write clock); and read from fifo_out on rclk The declaration has to create certain that any information was written on the FIFO at the write pointer, the equal records is read out from FIFO when study pointer is similar to the write pointer.

In this assertion, data_check property exams to see if FIFO is free. If so, saves wr_ptr into the nearby variable "ptr" and the information from FIFO into local variable "data" and show, so that we can observe the announcement is developing for the duration of simulation.

If the forerunner is true, the resultant says that the first match of rd_ptr being the identical as wr_ptr and that the read records is the same as the write records.

Sequence rd_detect(ptr) is utilized as a representation to first_match. It claims that wait from now till continually until you discover a read, and its rd_ptr is similar to the wr_ptr (which is saved in the local variable "ptr" in the antecedent).

Numerous assertions can be written to observe the operations of synchronizer graph. For example, strive to write simple assertions for your synchronizer design [7].

8.4   GALS STYLE OF COMMUNICATION

The communication method in NoC obeys the GALS fashion through the usage of a dual-clock FIFO in every data forwarding device [12]. Each data forwarding device in the community has a separate clock (e4, rd-clkl for data forwarding device-1, rd-clk2 for data forwarding device-2, etc.), as proven in Figure 8.8. As NoC helps mesochronous clocking, these clock frequencies have been assumed to be the same, whereas phases may additionally differ. The enter FIFO of data forwarding device-2 sends a request sign (i#-req2 = 1) to data forwarding device-1 till it is full. Data forwarding device-1, after receiving the request signal (in-reg2), sends 32-bit records (ffif2) and a records valid signal (in-val2) to data forwarding device-2 which is synchronous with data forwarding device-1's very own clock (rd-clk1). Data forwarding device-1 also sends this clack signal to data forwarding device-2, which uses it as the write clock of it enter FIFO (wr-clk2). The in-reg2 sign is synchronous with wr-clk2, which is the same as rd-clk1. Therefore, all the indicators (in-req2, sequence of data that is transported across a network2, and in-val2) between data forwarding device-1 and data forwarding device-2 in Figure 8.8 are synchronous with rd-clk1. Similarly, in-req3, sequence of data that is transported across a network3, and in-val3 signals are synchronous with rd-clk2. As every data forwarding device makes use of a separate neighborhood clock and these clocks are globally unbiased of cache different during the network, this communication method leads to the GALS style [12].

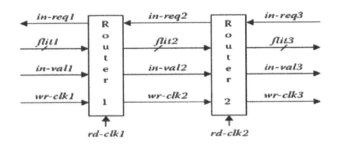

FIGURE 8.8   GALS communication.

8.5   WORMHOLE ROUTING

Space Wire routing switches appoint wormhole routing [11]. When a block of data starts to arrive at an enter port of a data forwarding device, its target tackle is looked at immediately. If the output port that is to be used to forward the block of data in the direction of its vacation spot is no longer currently being used, the head of the block of data is dispatched to that output port straightaway, with the rest of the block of data following as it is obtained at the input port. There is no buffering of an entire block of data in the data forwarding device, neither before, nor after exchanging [16].

Wormhole routing has a range of blessings over other procedures like shop and forward:

- Lack of block of data buffering;
- Small buffer memory;
- Can guide block of data of arbitrary size;
- Rapid exchanging.

Wormhole routing has one most important issue, that of blocking. When the output port that the block of data is to be transmitted via is no longer prepared or is presently being utilized, the block of data has to stay until it is ready or the block of data presently moving via it has completed. As the tail of a block of data can be unfold out via the network [1, 15], not only is the waiting block of data stopped, but that block of data stops any other block of data in the community which is ready to utilize the links that it is presently engaging. This is explained in Figure 8.9.

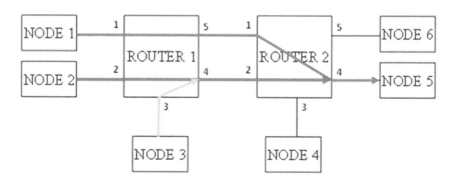

FIGURE 8.9   Block of data blocking.

A lengthy block of data is being transmitted from node 1 to node 5, displayed in blue. Another block of data, displayed in red, desires to go from node 2 to node 5, however when you consider that the hyperlink from data forwarding device 2 to node 5 is previously in use, the crimson block of data is stopped up in data forwarding device 2. A 0.33 block of data, shown in yellow, desires to go from node 3 to node 6. This does not use any of the hyperlinks being used with the aid of the first block of data (blue), but it is stopped through the waiting block of data (red) in data forwarding device 1 considering it has to tour over a hyperlink from data forwarding device 1 to data forwarding device 2. Once there is a blockage its consequence can multiply, producing similarly blockage at some stage in a network [16]. Blocking can appear for a number of purposes:

- A huge block of data is being transmitted;
- The target of the block of data is not prepared to get hold of it;
- Something has long past incorrect on the network, e.g., a connection failure, so that a block of data cannot go forward throughout the unsuccessful connection.

There are few methods that make easier to prevent blocking a network:

- Giant block of data is divided up into several tiny ones, e.g., a photo could be dispatched as a collection of photograph lines;
- Make certain that the target is prepared earlier than sending the block of data, which can be executed the usage of an end-to-end glide control method;
- If the target is no longer equipped to get hold of a block of data it can genuinely throw the block of data away, this can be shared with a retry method to enforce glide monitor, although it may end producing inefficient use of network bandwidth if the vacation spot is frequently now not ready;
- If a block of data does get stopped up for longer than maybe supposed, denoting that a fault has happened, notice this usage of a watchdog timer, and release the jammed block of data.

8.6 FUNDAMENTAL CHANNEL (VC) DATA FORWARDING DEVICE ARCHITECTURE DESIGN

Data forwarding device structure is a main location in NoC. To handle the hardware IP cores in multiprocessor system on chip (SoC) [2–4], it wants an extra help known as NoC. NoC can take control of the amount of IP cores by means of data forwarding device structure and the algorithms based on routing [3]. There are a variety of available data forwarding device configurations present such as wormhole and look-ahead data forwarding device architecture [3]. No buffering gadget is present in wormhole design [7] and hence it acts as more complex. This complexity can be determined by modifying with Fundamental paths to every enter port. A Diagonal Fundamental Channel data forwarding device is suggested, which has six directional ports, and one of them is dedicated to neighborhood port and balance ports are East, West, North, South, and diagonal ports. The fundamental channel (VC) data forwarding device plan is shown in Figure 8.10.

The VC data forwarding device assists to stay away from the trouble of site visitors traffic and routing delay in the group. The objective of the VC data forwarding device is to produce the bandwidth effectivity and improved throughput. This arrangement has more characteristics integrated as like low delay, monitoring the site visitors traffic, and buffer the block of data depend on risky storage method. The first move in the digital channel data forwarding device is to create a record into the buffer and it is known to as buffer queue (QW). In the 2nd step, the routing calculation (RC) unit allocates the target address to the header sequence of data that is transported across a network [1, 14]. Then, depending on the header sequence of data that is transported across network information, the block of data is routed to the target. In reality, Switch allocator [11] and crossbar network video display units the nearby data forwarding device to take a look at whether they are engaged or open. With this, data forwarding device will determine to transfer the block of data to nearby data forwarding devices block of data. suppose, if it is busy, then it will stall the block of data in digital paths. In the 1/3 step, VC assigns the block of data, depend totally upon the delivers obtained by means of the swap allocator (SA) [16].

SA can restore the completed indication provided to the VC and VIP allocators. Depending on the grants, VC allocator assigns the block of data into the buffer queue. If the buffers are full, then it shutdowns the block of data; or else, it passes the block of data to the cross-bar network. In the fourth step, swap traversal (ST) transmits the block of data to the related output port and hyperlink traversal (LT) is associated to the equivalent succeeding data forwarding device. Switch traversal (ST) [11] and LT are part of VIP. This data forwarding device will function primarily depending on the size dispensed routing algorithm. A small transform is carried out in VC data forwarding device layout by means of inserting a new diagonal port to the muter [2, 13].

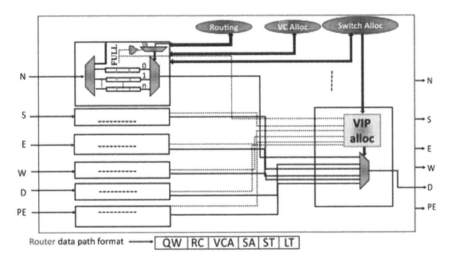

FIGURE 8.10   Fundamental channel data forwarding device design.

8.7   CONCLUSION

A hybrid method depending on virtual circuit exchanging is proposed to decrease the delay occurring before the data transfer and power of NoCs. The fundamental thought of the suggested hybrid method is to execute virtual circuit exchanging beside with circuit exchanging (CS) and blocks of data exchanging (PS). A shortest route algorithm is also offered to smartly assign VCS connections and CS connections for a specified traffic in mesh linked NoCs, so that the time consumed to transfer the data in blocks. The delay occurring before the data transfer and total energy utilized by the NoC circuit is minimized. Blocks of data based NoC structures are presently the leading selection to discuss the data transfer needs of SoCs. The resource-controlled nature of such networks distinguishes them from conventional off-networks, and hence, requires exact and effective analysis of their presentation, fault-tolerance, and energy activities. We suggest a queuing-theory-depend systematic model for 2D mesh networks, which executes delay occurring before the data transfer and power testing at the quality of special router sub-blocks for improved precision.

KEYWORDS

- **asynchronous FIFO**
- **block of data**
- **circuit exchange**
- **globally asynchronous locally synchronous (GALS)**
- **routing techniques**
- **virtual cut through (VCT)**
- **wormhole (WH) exchange**

REFERENCES

1. Paramasivam, (2015). Network-on-chip and its research challenges. *ICTACT Journal On Microelectronics, 1*(2).
2. Ahmed, B. A., & Slim, B. S., (2013). A survey of network-on-chip tools. *International Journal of Advanced Computer Science and Applications, 4*(9).
3. Faudzi, A. A. M., & Suzumori, K., (2010). Programmable system on chip distributed communication and control approach for human adaptive mechanical system. *Journal of Computer Science, 6*(8), 852–861.
4. Ankur, A., Cyril, I., & Ravi, S., (2009). Survey of network on chip (NoC) architectures & contributions. *Journal of Engineering, Computing, Architecture, 3*(1).
5. Ahmad, B., Ahmet, T. E., & Sami, K., (2006). Architecture of a dynamically reconfigurable NoC for adaptive reconfigurable MPSoC. *Proceedings of the First NASA/ESA Conference on Adaptive Hardware and Systems.* IEEE Computer Society.
6. Helali, A., Soudani, A., Bhar, J., & Nasri, S., (2006). Study of network on chip resources allocation for QoS management. *Journal of Computer Science, 2*(10).
7. Benini, L., & Bertozzi, D., (20050). network-on-chip architectures and design methods. *IEEE Computers and Digital Techniques, 152*(2), 272.
8. Pratim, P. P., Grecu, C., Jones, M., Ivanov, A., & Saleh, R., (2005). Performance evaluation and design tradeoffs for network-on-chip interconnect architectures. *IEEE Transactions on Computers, 54*(8), 1040.
9. Wolkotte, P. T., Smit, G. J. M. Rauwerda, G. K., & Smit, L. T., (2005). An energy-efficient reconfigurable circuit-switched network-on-chip. *Proc. 19th IEEE International Conference on Parallel and Distributed Processing Symposium* (pp. 155–163).
10. Kim, D., Manho, K., & Sobelman, G. E., (2004). CDMA-based NoC architecture. *Proc. IEEE Conference on Circuits and Systems, 1*, 137–140.
11. driahantenaina, A., Charlery, H., Greiner, A., Mortiez, L., & Zeferino, C. A., (2005). SPIN: A scalable, packet-switched, on-chip micro-network. *Proc. IEEE Conference on Design, Automation and Test, 70–73.*
12. Nikolay, K., & Gerard, S. J. M., (2005). *A Survey of Efficient On-Chip Communications for SoC.* Department of EEMCS, University of Twente, Netherland.

13. Rickard, H., & Magnus, H., (2002). *Modeling and Prototyping of a Network on Chip.* Master of Science Thesis, Ingenjorshogskolan.

14. Jantsch, J. S., Forsell, M., Zheng, L., Kumar, S., Millberg, M., & Oberg, J., (2001). Networks on chip. *Workshop at the European Solid-State Circuits Conference.*

15. Hemani, A. J., Kumar, S., Postula, A., Oberg, J., Millberg, M., & Lindqvist, D., (2000). Network on Chip: An architecture for billion transistor era. *Proc. of the IEEE Nor Chip Conference.*

16. Ashok, K. N., & Kavitha, A., (2016). A novel 3D NoC scheme for high throughput unicast and multicast routing protocols. *Tehnički Vjesnik, 23*(1), 215–219.

CHAPTER 9

Routing Strategy: Network-on-Chip Architectures

N. ASHOK KUMAR,[1] S. VISHNU PRIYAN,[2] P. VENKATRAMANA,[3] and DURGESH NANDAN[4]

[1]*Department of Electronics and Communication Engineering, Sree Vidyanikethan Engineering College, Tirupati, Andhra Pradesh, India, E-mail: ashoknoc@gmail.com*

[2]*Kingston Engineering College, Vellore, Tamil Nadu, India*

[3]*Sree Vidyanikethan Engineering College, Tirupati, Andhra Pradesh, India*

[4]*Aditya Engineering College, Surampalem, Andhra Pradesh, India*

ABSTRACT

Network-on-chip (NoC) is a special and unique case of parallel computing systems defined by tight constraints such as availability of resources, area, cost of the NoC architecture and power consumption. A NoC is designed of three main components: switches, network interfaces (NIs), and links. It is a technology that is proposed to solve the shortcoming of buses. This technique is to design communication subsystem among IP cores (intellectual property core) in a SoC design. The strategy by which bundles are guided from source to destination can be deterministic, unaware, or versatile. Systems with no way differences are compelled to utilize deterministic directing techniques; different topologies may utilize unaware or versatile calculations. The three important things to design NoC is topology, routing mechanism and switching algorithm. The topology defines how to interconnect the nodes in the network and switch is enabled after receiving the header of the packet that contains the information about the destination address. The switch input channel is connected to the output channel. There are many switching techniques used: store- and forward switching, cut-through switching, and wormhole switching. The routing algorithm and network topology are the two important aspects for on-chip communication in NoC. The router makes the path for the received packet to the destination. The main property of route in NoC should make the path to error-free, deadlock-free, and highest performance. The route in the NoC is divided into two parts, one is the source node which makes the path before packet leave from the source, and another one is the destination route which make the path based on the information given in the header of the packet.

9.1 PACKET ROUTING

In the network of packet switching, routing will be the choice that is top-level that coordinates arrange packets from their source towards their location through center system nodes through particular packet practices that are forwarding. Packet forwarding is the transfer of network packets from the community that is solitary to another. Every bundle comprises three parts that can be particular each right component has a portion associated with the portions as follows above: Header: frequently includes an alarm signal showing that the data are sent, supply, and target forwarding and clock information to synchronize the transmission.

9.2 QUALITY OF SERVICE (QOS)

Quality-of-service (QoS) can be a choice of recent available technologies, which controls bandwidth utilize whenever information surpasses network of computers. Its many applications that is basically created for the project of real-time with a high precedence information application. QoS technologies has specific rolls used in combination with another in order to make a system that is end-to-end policy. Categorization alongside queuing will be the two generally used QoS tools with regards to manage traffic. Categorization recognizes with evaluates traffic towards make sure community items identify exactly how regarding the way for acknowledge afterwards prioritize information as it proceeds through something. Queues are safeguards within tools along with the intent behind retain data designate ready. Queues offer bandwidth requirement alongside prioritization of traffic because it inserts otherwise eliminates a grouped community device.

Condition of the queues are not empty; they disseminate with decrease traffic. Control along with smoothing moreover generally used QoS technologies to control the bandwidth used in executively described traffic categories. Procedure insists on bandwidth towards a particular range. But relevancies attempt to utilize additional bandwidth than the allocated one, their traffic desire to be the located observation otherwise reduced. Construction to describe software put maximum resting on the transmission rate of bandwidth used for a crew of information data. But additional data traffic is required to be real transfer than the twisted range permits, the surplus resolve be really protected. This protect be capable of followed by use queuing towards prioritize data when it leaves behind the protect [13]. Totally, link-particular product moreover compression tools are employed by reduced bandwidth WANs near make sure live applications do not interfere the higher-end data with interruption. The WRED (weighted discard that is random early technology) delivers a congestion avoidance method which decreases lower concern TCP information towards test ways to protect greater concern information starting the adversative after-effects of congestion [12].

9.3   PACKET FLOW THROUGH A TYPICAL QOS POLICY

9.3.1   QOS DEPLOYMENT LIFECYCLE

The measures being subsequent a QoS consumption lifecycle:

1. Development preparation with obtain into understand present in addition to nearby future QoS demands concerning the relationship just because an over that is entire above useful for all responsibility. Decide on a QoS that is appropriate representation obtain responsible obtain within early in the day you begin [13].

2. **Review and Plan:** If you are building gear that is extremely important PC software improvements, build entity is initial. Then:
   i. Snapshot current policies which can be QoS you want near rollback;
   ii. Examine the QoS convenience of your community items;
   iii. Baseline the system through problem viewing and workout use of testing;
   iv. pick a QoS type useful for the traffic classes you may want to encouragement;

v.   Identify QoS instructions helpful for center of operations and home LANs;

vi. Identify QoS directions employed for WAN links and spot that is domain.

3. **Evidence of Concept (POC):** Review QoS guidelines and backgrounds initial in only a non-construction location with genuine and traffic that is unreal produce handled circumstances. Analysis separately useful for the insurance policy and followed by for instance the instructions being entire.

4. **Iterative Usage Rotation:** Compress QoS policies within a segmented technique, each near portion regarding the system otherwise through QoS energy (categorization then queuing). Approve the alterations close to all iterations ideal for minimal a day early in the time progressing nearby the style that is next.

5. **In Progress Monitoring and Assessment:** Entire In progress monitoring and correct your polices not now used by standard for a foundation that is day-to-day although furthermore in help of month-to-month, quarterly, and establishment that is annual.

9.3.2 QOS CONSTRUCTION WORKFLOW

Apply quality of service (QoS) policy in direction of boundary (Figures 9.1(a) and (b)).

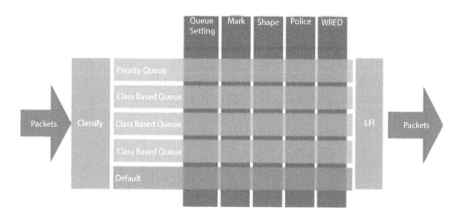

FIGURE 9.1(A)   Packet flow through a typical QoS policy.

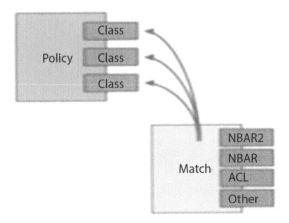

FIGURE 9.1(B)   QoS policy.

Period 2: Traffic decides to go there because it has been given a choice that is the first QoS policy next to traversing interfaces.

9.3.3   QOS CONFIGURATION WORKFLOW (FIGURE 9.1(C))

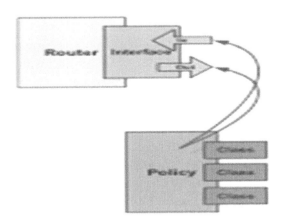

FIGURE 9.1(C)   QoS workflow.

Phase 3—Stage 1: Describe traffic classes and set you back QoS policy (Figure 9.1(d)).

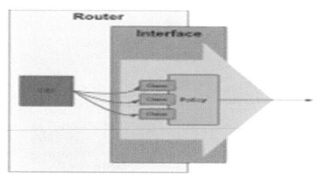

FIGURE 9.1(D)   QoS traffic classes policy.

9.3.4   QOS TOOLS

Only a little is talked about by us regarding the necessity of QoS along with different application category containing various requirements. Presently permission to discuss just about the apparatus which may be genuine may use to work well with QoS:

1. **Categorization and Marking:** But you want to provide dependable packets an action that varies we must recognize and correct them.
2. **Policing and Determining:** Those two tools are used to speed-take control of your traffic.
3. **Queuing-Congestion Managing:** As a substitute of getting a particular queue that is big packets are consider with FIFO, we could create a few queues with different priorities.
4. **Congestion Avoidance:** You shall find chosen tools used to carry out packet loss and near decrease congestion [15].

9.3.5   CLASSIFICATION AND MARKING

Categorization practices—there are lots of methods for categorizing traffic. A few procedures that is familiar:

- Entry Monitor listing executed to a locate-map (IP target, port, etc.);
- Variety of Application executed in the direction of the acknowledgment map (through NBAR or NBAR2);
- Control port assume state.

A packet is initially recognized, and it is usually marked with the traffic problem associated to a source, making of packets is the best way to get make an action that is somewhat different. Correct is normally continued during the Layer 3 IP header by means of the services that is differentiated code (DSCP) [11]; LAN executions use the course of service (COS) domain.

9.3.6   QUEUING

Numerous type of queues are designed into community products. Generally, class-based weighted fair queuing (CBWFQ) and priority queues (PQs) are the two general categories of Cisco products. PQs are thought for use with packets which calls for lesser latency and management that is least-jitter. PQs want decrease several ended contributions for the duration of the period of congestion. CBWFQ is called useful for mass alongside traffic in a transactional manner which are not the same as jitter-precision or period loss. Each queue can indicate an engaged with the bandwidth that is sweared to be presented made for usage by that set-in extent of obstruction. At last, queue deepness may be fixed to traffic that is certified can be controlled with a queue [1].

9.3.7   SHAPING AND POLICING

Policing includes assessment the packet cost developed for the categorization that is for certain of to find out if it trusts to price that is sure. The policy determination promises to see condition there is traffic that is huge and condition traffic that is excess, it could reduce the packets in arrange to concur with to that particular rate that is certain.

9.3.8   WAN TRAFFIC SHAPING

Designed for WAN links, we are able to locate backgrounds specially meant for aggregators of WAN that contain shaping, marking, queuing, policing, and URL division. In numerous possessions, there command to be present a connect-speed variance involving town that is interior slower outside WAN link. Use policies being hierarchical contour the QoS policy to the WAN linkability.

9.4   CONGESTION CONTROL FOR NETWORK ON CHIP

Primarily NOC execution is founded at networks in a lossless domain consists of much little, safeguard modules with switches (SMs or routers) [14]. Including safeguard within every SM avoids packet loss in the event of disagreement but establish extra delays that are queuing. Small buffer restriction decays the technique operation in conditions of latency and throughput when traffic pack is heavy, and make problems the style of obstruction manage method [13]. Congestion may appear next to the SMs, so called intersection congestion, or at heavy loaded endpoint, so named hotspot. Lower rate of routing in the Mesh that is 2D and networks may result in link obstruction [15]. Figure 9.2(a) indicates the identification of problem that is nagging of link obstruction in 2D Mesh [1, 8]. As a result of the half section that is low and routing that is reduced website link congestion simply take place and decay is the network execution. Numerous endeavors have actually exposed that adaptive routing improves website link congestion by steering traffic perhaps not here beginning the links that are congested [10]. Conversely, adaptive routing would weaken unchanging additional connections in the community whenever a hotspot initiates because exposed in Figure 9.2(b). Planned the instant that is last, the high bisection bandwidth and symmetric description of Clos NOC (CNOC, see Figure 9.2(c)) allow traffic to be regularly scattered between the SMs in the phase that is middle. Via a good routing that is load-balanced, CNOC have actually to experience less link congestion than the beyond specific networks.

Conversely, the congestion that is source hotspot trouble at rest be real. When a hotspot occurs, it begins approaching back once again the traffic beginning resultant that is upstream SMs head-of-line (HOL) jamming in to the SMs. The so-called sensation is termed obstruction that is saturation-tree as exposed in Figure 9.2(c), where initial the congestion instant (e.g., the hotspot PE in Figure 9.2(c)), the traffic moves specific for the congest node create a tree and will not shift redirect. Packets destined for low loaded objectives can also show up jammed by the jammed traffic in the interior SMs and experience at the time of high delay that is queuing. Our presentation research illustrates that throughput of CNOC's decay is certainly below the traffic in the hotspot because exposed in Figure 9.3.

Figure 9.2 describes obstruction struggles in CNOC and 2D [15] (a) link obstruction in 2D mesh. PEs linked to point nodes 1, 2 and 3 traffic that are ahead of PE linked to point node B along with the routing of X-Y. Node X is obstructed because of website link obstruction just starts as soon as

the traffic enters next to the location. (b) Hotspot obstruction in 2D Mesh
[15]. Usage of adaptive routing traffic supported in the feasible dashed lines;
several nodes (term X) dedication be here congested together with issue gets
of poorer quality. (c) CNOC Hotspot congestion. The long half-section bandwidth and balance loaded routing to improve URL obstruction in CNOC; in
saturation-tree traffic in hotspot founded as an issue [7, 8].

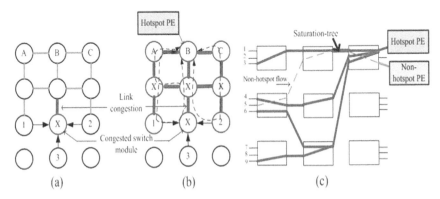

FIGURE 9.2   Obstruction struggle in CNOC.

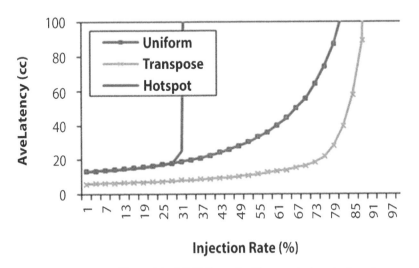

FIGURE 9.3   Various traffic model of CNOC standard latency during clock cycles.

Hotspot obstruction monitor is among the issue that is generally difficult
whenever scheming is increased-throughput lesser-latency NoC. Just to

recognize the incidence how its behavior(s) of hotspots and inform every resource just before control their traffic towards the hotspot(s) is really complicated considering the fact that related to possibly high-pitched capability of in sequence towards be present composed as the latency that is unimportant the recognition obstruction node along with the input. Using this endeavor, we suggest a point to point that is efficient managing system, termed as HOPE (hotspot prevention), to look for the hotspot obstruction downside for all CNOC and 2D Mesh [1]. The introduced traffic percentage is practically with approximating the packets numbers included by the community designed for all target and utilized as an effortless stop-and-go protocol to end traffic of hotspot starting the internal links jamming for any system in certain, HOPE control. We estimate HOPE's generally talking implementation and its scalability and equality. Wide-ranging reproduction outcomes cantered on altered traffic models verify that HOPE can get better method effectively execution by moderately hotspot that is changeable. Our hardware assessment illustrates that HOPE features a logic that is extremely little.

9.5   FLOW CONTROL IN NOC

9.5.1   FLOW CONTROL

1.   The routing of the message needs circulation of varied resources: the path (or link), buffers, control state.
2.   Buffer less: darts are fall condition there is conflict intended for a hyperlink; NACKs are delivered right back, together with transmitter that is initial to re-transmit the packet [13].
3.   Circuit switching: a need is initially delivered to protect the businesses, the requirement may be expected at a router that is middle the machine is accessible (ergo, perhaps maybe not properly buffer-less) [12], ACKs are delivered back, and following packets/darts are routed including small effort (good intended for mass transfers) [2, 5]:
     i.   Flow control make a decision exactly how the possessions associated with neighborhood as an example neighborhood buffer as well as transmission capacity cap capability are corrected to packets passing through a domestic area;

    ii.   Goal is to use resources as effective as feasible to tolerate a throughput that is high;

   iii.   A capable motion control is merely a requirement to get a neighborhood execution that is great;

   iv.   Flow control can be deemed an issue that is nagging of:

        a.   Resource allocation;

        b.   Contention quality.

Resources in type of websites, barriers as well as problem need to be right here allotted to every package. If two packages enter for any type of area that is equivalent control can just assign the neighborhood to 1 packet, but demand furthermore handle one other packet.

Flow control can be divided right into:

- Buffer less circulation control: Packages often tend to be moreover misrouted or family tree;
- Buffered circulation control: Packets that cannot be directed using the preferred system are conserved in barriers.

9.5.2   CIRCUIT SWITCHING

Circuit-switching is in fact an activity that is bufferless, where an amount of systems is required to describe a path [5]:

- A need (R) circulates supply that is starting location that is replied by the recognition (A).
- When information is sent (the adhering to two five-info packets (D)) as well as end data (T) is offered reallocate the business.
- Circuit-switching does not require lower packets being or else misrouting but there are two primary constraints.

High latency: T0 = 3 H is r + L/b wire that is (neglecting).

- Low throughput, considering that system might be used up to the component that is big or made use of by indicating instead of for transfer for the haul [12].

9.5.3 FLOW CONTROL

Better resourceful circulation control can be reached by including buffers by enough barrier's packets do not call for to be really misrouted or reduction, given that packages can be on hold for the outbound network to be prepared [2, 13].

The two main approaches are:
- Packet-buffer flow control:
  - Store-and-forward;
  - Cut-through.
- Flit-buffer flow control:
  - Wormhole flow control;
  - Virtual channel flow control.

9.5.4 DATA UNIT

1. **Information Unit:** Pictorial representation of message unit is given in Figure 9.4.

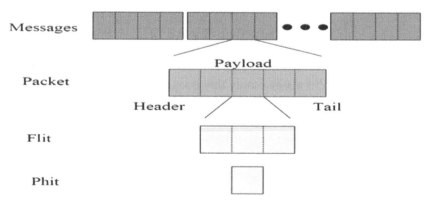

FIGURE 9.4 Message unit.

2. **Store and Forward Flow Control:** Every factor along a course holds to a package is totally get with your package is sent towards the factor that is following. Two sources are anticipated:
   - Pocket-sized barrier in the button;
   - Exclusive usage of the channel that is outgoing.

3. **Cut-Through Circulation Control:** Transmission on the following networks starts straight that is right the latest header dart is gotten while the sites are released after tail dart T0 = H t r + L/b 12.

4. **Drawbacks:**
    - Exceedingly inefficient use of barrier area;
    - Buffers are allocated in devices of packages;
    - Often, we require a few barriers that is self-determining to diminish jamming before offer predicament avoidance contention latency is boosted;
    - By assigning buffers within entity of packets improved, e.g., critical package have a crash with a low-priority packet;
    - Must wait the low-priority that is total to be transferred in advance it might have the community.

5. **Wormhole Flow Control:**
    - Wormhole flow control function similar to cut-through, apart from surrounded by system as well as buffers appointed to darts rather than packets [10].
    - Three resources are expected:
        o   A digital neighborhood for any packet: Body flits of a packet take advantage of the VC gotten through the slit that is starting;
        o   One flit buffer; and
        o   One flit network bandwidth.

6. Virtual Channel Flow Control:
    - Each switch has a few sites which are virtual. A network is a group of networks that are virtual;
    - Each network has an output network that is assigned to the head, and sends buffered packets to the other networks;
    - A mind flit must designate the identical three resources through the switch that is next being forwarded;
    - Because of containing many virtual companies for each system that is real two various packets are approved to use the channel and not waste the resource when one packet remains perhaps not used.

9.6   ROUTER DESIGN

NoC system is representation that is genuine a layout anywhere within nodes, processing elements (PEs) as well as sides will be the connective links related to the PEs. Figure 9.5 highlights the NoC that is essential design it essentially contains processing aspect (PE), router [14]. Each PE is gotten in touch with NI which connects the PE to a router that is regional. Each time a packet finished being provided starting a supply PE approximately an area PE, the package is forwarded jump by get on the system using the alternative made up of all router. Estimating within practically any system, router is one of the most elements that are necessary for the design of transmission strength of the NoC system. The capability for any kind of router is always to reward a gotten package towards the location source inside a packet-switched system. Condition is properly affixed to it, otherwise towards forward the packet to some other router connected to it [5, 8]. It is significant, that the layout of the NoC router ought to be as practical and also simple since execution price boosts having a boost into the appearance intricacy regarding the router [5]. Transmitting techniques helpful for neighborhood pick the route developed for details transfer in the direction of the area point. The synopsis of a particular method is listed here.

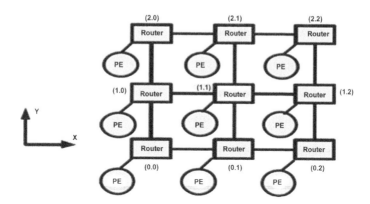

FIGURE 9.5   Router design.

9.6.1 XY ROUTING ALGORITHM

Transmitting formula might be made use of two courses the packages supply that is beginning to place PE. XY transmitting formula is made use of path packet during proposed router making. This is a selection of distributed routing that is deterministic. All routers can be acknowledged through its take care of as exposed in Figure 9.5 in 2-D mesh [9, 15] NoC. The XY transmitting algorithm evaluates the router that exists (Cx, Cy) to the area router target (Dx, Dy) of the packet, kept within the header flit. Flits has to be directed in the direction of the core port connected with the router whenever (Cx, Cy) target of the router that exists equal to the (Dx, Dy) target. The Dx target is contrast well to initially the Cx (straight) address should this be not the instance. Flits commitment is right here routed to the East slot after Cx Dx and also when Cx= Ax the header flit is by now horizontally connected. However, this disorder that is last real, the Dy (vertical) address is a review to the Cy target. Flits will certainly be routed to Southern as CyDy. In situation chosen slot is busy, the header flit alongside completely after flits of this packet resolution be here obstructed. The transmitting request this package will stay working until a web link is acknowledged in several executions that is future of approach within this router.

9.6.2   SURROUNDING XY ROUTING

Surrounding XY directing (S-XY) has three transmitting that is various. N-XY (Regular XY) setting performs just such as the uncomplicated XY directing. Its roads packages first along x-axis and also afterwards close to y-axis. Routing keep NXY setting as long as the network is not jammed and also routing does not put together routers that can be inactive. SH-XY (border XY that is utilized when the router's left or ideal neighbor is handi-capped. Subsequent the setting that is third (envelop upright XY) is made use of after the greater or lower neighbor from the router is unused. The setting that is SH-XY packages to your accepted function throughout the premises to organize the location. The algorithm prevents packages pretty much the defective routers down the fastest course that can be done. The problem is simply a little bit that is little to the SV-XY mode for the primary reason that the packets are in fact into the ideal column. Packets are transmitted to left or suitable. The routers within the SH-XY and SV-XY modes insert a recognizer that is tiny the packages to inform in the direction of last routers why these packages are routed utilizing SH-XY or setting that is SV-XY. Hence, the routers which could possibly be last not send the packets towards the trunk. Nearby XY directing is employed inside a DyNoC. This may be a process that maintains transmission in between components which are dynamically situated having an item.

9.6.3   OE ROUTING ALGORITHM

OE routing formula is usually a dispersed directing that is adaptive, which is often supported on odd-even turn design. It has constraints that can be couple of meant for protecting against and also avoiding at the time of predicament development. Odd-even turn model smooth the development of deadlock-free transmitting in two-dimensional (2D) fits together through no organizations being essential. In the mesh that is proportions that are two-dimension every node is acknowledged through its synchronizes (x, y). A column is acknowledged as even when its x dimension aspect is furthermore mathematical column in this version. Furthermore, a line is referred to as weird problem it is a dimension aspect is most definitely an amount that is strange. A turn comes with a 90° change of traveling direction. A modification is just a 90° change in an identify with summary. There are eight types of turns, on the basis of the traveling standards for any kind of recurring companies that are attached. An adjustment is referred to as ES turn conditions a customize be had by it of instructions East that is beginning to. Furthermore, we are able to describe the final seven types of turns, particularly EN, WS, WN, SE, SW, NE, as well as NW transforms, where E, W, S, as well as N suggest East, West, South, and North, for the reason that acquisition. Whenever a full, there are two problems that are major.

- No packet is permitted to do EN turn in every node that will be are readily available on an additional column. Likewise, no packet is permitted in the direction of full NW renovation in many nodes to be put by the line that is weird; and
- No packet is allowed in the direction of complete ES enhancements in most nodes to even be inside a line. In addition, no packet is allowed towards complete SW modification in many nodes which remains in a loved one line that is strange.

9.7   DESIGNED NOC LINKS

Several designers are dependent upon simulation, when they believe to expect the efficiency of SOC in the starting phases of growth. In the eventuality of network-on-chip (NoC) style, two essential parameters show that self-simulation is not any longer suffice to supply a style that is maximized [12].

9.7.1 LINK EFFICIENCY MECHANISMS

LFI planned for MLP (link fragmentation and also interleaving for multi-link PPP) separates, re-joins, and also show datagrams across a couple of information links which are reasonable. Large delay-sensitive datagrams are multilink enveloped and also separated into little sized packets to wait that are meeting. Minimal packages which are typically delay-sensitive perhaps not encapsulated, but are interleaved linking these fragments. LFI made for frame relay (FRF.12) pieces packages that are huge then the locations you indicate using the piece size need that is frame-relay. Frame relay cannot differentiate among VoIP and information, build the fragmentation ergo range during the DLCI therefore speech frames are not split. The fragment range be mentioned within bytes (default = 53 bytes). Plenty of facets decide the true number related to speech packets. Compressed time that is real (CRTP) decreases within line expenses meant for multimedia RTP traffic resultant throughout a consequent decrease within delay. CRTP is in particular valuable while the RTP payload range be little for example compressed sound payloads of 20 to 50 bytes.

9.8 MULTICAST ROUTING SCHEMES

Multicast directing can be separated into tree-based as well as path-based algorithms.

9.8.1 TREE-BASED MULTICAST

The basis together with the entire area nodes specific the fallen leaves regarding the tree throughout a tree-based multicast routing algorithm the networks of the package range an on both sides of the tree via the structure node person. This makes sure up to and consisting of quickest course in the middle of your supply node and each location node is continuously taken. Boscage of packets might get in via every one previous to get into the production that is similar of routers, which settle boost area contention and also congestion then might result during predicaments right below details instances [6]. Figure 9.6 highlights an illustration of this a multicast deadlock, where two packages, especially Package and Package B, send an application for the western while the eastern manufacturing ports of Router 2. Presume Package A is setting up in the direction of make usage of the western outcome port except needs in the direction of delay employed for the eastern manufacturing slot to has been provided to Package B, while Package B is given to work with the eastern output slot but furthermore needs to utilize the production port that is western.

All packet willpower holds one production port as well as awaits the port that is next by the package that is final. A hold that is circular wait condition occurs as well as both packets cannot go additionally, resultant in the multicast deadlock.

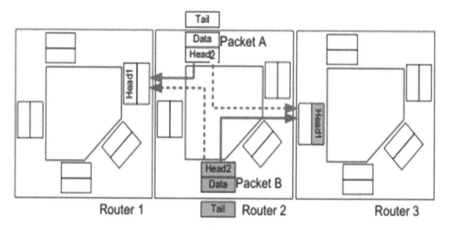

FIGURE 9.6   Deadlock problem of multicast routing.

The hardware tree-based multicast routing algorithm (HTA) understands a deadlock queue within all routers during purchase to deal by the deadlock drawback. Condition a mind flit cannot be routed towards a grouped community in to a determined amount of time, the pinnacle flit is determined in way of happen as being a deadlock condition and provided for the deadlock queue connected with node that is current. Ergo, the flits that are residual with packet can ahead be communicated without any longer jammed aided by the head flit, therefore releasing the deadlock. Every time a determined amount of time, the packet within the queue will undoubtedly be re-injected into the input [14] connected with a router that is present. Hence this packet wants be there communicated after the before jammed packet.

The origins suggest a router model to help keep away from deadlocks by way of a selection of six tips ideal for an input buffer. Every input has actually become greater towards that are sufficient a packet that is entire [14]. A pointer is included by the pointers indicating near the flit that is initial the buffer moreover every one of the final five pointers showing in direction of the flit become transmitted to 1 output slot next. Condition a packet is jammed due to the flit that is initial to the buffer; the router can continue transmitting the flit to is pointed through the pointer ideal for a few production ports that is non-blocked. Showing up in this technique the deadlock problem is fixed.

9.8.2   PATH-BASED MULTICAST

Path-based multicast directing formulas steer clear of predicaments with threatening split to routers as well as this can be numerous. These algorithms merely replicate a package in the direction of the restricted manufacturing port although the packet gets in by the person of its locations. All nodes (x, y) in just a letter by means of a 2-D mesh are designated making use of a tag L(x, y) in this procedure [6]. Packages are sent out according in the direction of the identify in to the technique of all package is transmitted near its neighbor router via a tag in between your router that takes place the location router. The areas of the packet are partitioned into two groups: locations through tags smaller sized and also bigger than the foundation node. The peak sweeps in all team that is combined split in the type to they are to be below disperse. In routing, each flits check your face that is initial or else the very head that is first access its place. The flit concurrently by duplicate related to the entire data flits plus completion flit are sending to the minimal node although the very first mind flit gain access to its location.

The head that is primarily right into the mind that is recurring adhered to closely by become the existing initial brain flit plus the packet maintain in the direction of keep for this really first head flit looking forward to the present actually initial mind flit gain access to its location. This method maintains eagerly anticipating every flit complete for their areas. Demonstrably by the path-based algorithms, a packet dedication head to every location below the equivalent line by directing towards other locations as being a line that is various. The benefit of these algorithms must be to deadlock that is negative look [6]. The disadvantage is the period of courses in the direction of cover-up every area node could be here expanded. Specific earlier algorithms which are path-based reduce the lengths of courses are future. The multi-path (MP) algorithm separations the area nodes awake in the direction placed out of joint established near reduce the roadway lengths. This algorithm is working into which other uses a directing that is adaptive increase the opportunity for choosing the non-congested course subsequently since to expand reduce the course lengths.

9.8.3   PATH-BASED MULTICAST ROUTING FOR 2D

NoC is a functional relationship that is adaptable taken into consideration developed for SoC. NoC totally depends in advance the transmitting algorithm utilized by its execution into transmission applications. NoC is absolutely a variant that is alternative of on chip that is implemented in the kind of network, so named as NoC, Theon chip transfer inside NoC will obtain set up because of package switching. Transmitting algorithm determine right into which program the packet may be sent out from supply to place, used by this work routers are utilized connecting nodes into the system, these routers want via the packets based in advance the algorithm that is directing within the type of NoC. The routing algorithm for the considered NoC leads to the strategy. Latency, load, and throughput sharing are specially constraint that is valuable be there calculated though producing [6] principle towards NoC. Some transmission method overall corresponding going on solitary chip is called NoC [12]. It is basically IP address included simultaneously at solitary chip among clock logics that could be asynchronous or synchronous. NOC since well workings at un-clocked logics furthermore.

This improves within improving the transmission on chip more accu-rately than just what a coach or perhaps a crossbar control resolve work through. In this, NoC improve the flexibleness of SoCs. NoC is a person who is frequently must be worldview that is significant for doing transmission between different centers in just a SoC. One of many enhance during IC manufacturing a reliable make an effort is to fabricate enormous computes of methods continuing the chip that is similar comprehend additional resourceful and improved poker chips. More a competent computation that is straight makes a difference near boost the effectiveness associated with routine surrounded within the chips. Ordinarily, SoCs work topologies taking into description provided buses. NoC is truly a quality that is capable the way of resolving the commitment of complicated design relating through on chip interaction. NoC is another worldview used for SoC. During NoC worldview, centers are linked through each other throughout a functional system of settings and so they communicate between by themselves during packets support switching.

9.8.4 ARCHITECTURE OF NOC

Mostly, inside NOC architecture, the chip that is entire portioned into a couple of interconnected blocks called as nodes. Each particularized node determination has PE that can easily be one of the signal processors that is electronic. All nodes in to the NoC can be embedded, including a router and also this router improves created for developing the connectivity between all nearest nodes of this community. Figure 9.7 defines the description that is illustrative of Node. A node is composed of four ports. The ports are West, North, East, and South; they are connected by past routers. There is certainly furthermore a port that is regional is linked towards the PE. Thus, totally you shall find five inputs and five outputs.

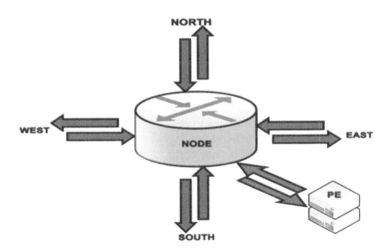

FIGURE 9.7   Illustration of node.

9.8.5 DESIGN OF NOC

The NoC and this can be created consists of 4×4 nodes understand predicated on Mesh topology [9], as considered before a node that is solitary a grouped community includes north, east, south, and west ports along through local in and local away. The input and production made for a node depends ahead the genuine number of the node which will be give explanation more (Figure 9.8).

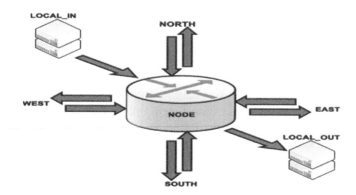

FIGURE 9.8   Illustration of a single node.

Basically, in direction of acknowledge the functioning of NoC, undertake us evaluate the execution of the node that is solitary. Made for a node, the inputs might be starting all the four instructions at by means of neighborhood input. In direction of acknowledge with an improving technique, undertake us evaluate a system that is 2×2. A community that is 2×2 exposed in Figure 9.9. A 2×2 mesh network comprises of four nodes N1, N2, N3, and N4. The inputs for N1 could be on east and south, N2 on east and, N3 being North, West and South and N4 on West and North.

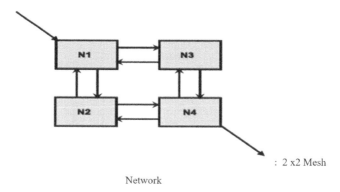

: 2 x2 Mesh

Network

FIGURE 9.9   2 × 2 mesh network.

It is actually measured towards take place Node-1, created for some node at hand resolve be there a clock and reset, the input packet information is of 48 bit and production packet of data is of 32 bits. At N1 designed for the

2×2 network, the feasible inputs are on East and South while we consider a specific node, the likely inputs and manufacturing are compared. Figure 9.10 illustrates the terminals for the solitary node, now. Therefore, decay after holds are inputs which mean manufacturing on N1 are there on East otherwise South.

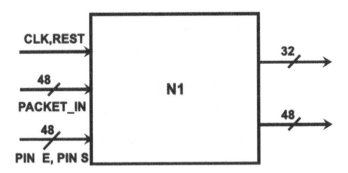

FIGURE 9.10 Terminals of a single node.

Figure 9.10 illustrate the output that is normal node1, whenever convenient is no input towards the node, there is not any output regarding the node, except while a packet of information take place on regional in the output be here similar to the input. As soon as a packet of data arises on east, the production is moreover on east. Also, packet by south occurs on just (Table 9.1).

TABLE 9.1 Analysis of Packet Out for Node 1

Pins	Pine	Packet-In	Packet-Out
0	0	0	0
0	0	1	Packet-in
0	1	0	Pine
1	0	0	Pins

9.8.6   ROUTING ALGORITHM FOR 2×2 NETWORKS

During XY routing algorithm, the packet which includes to be delivered more the system contains header flit, end flit and information flit. Destination node target be occurring the header flit. The motion chart of 2D routing algorithm is exposed in Figure 9.11. The target that is current (Cur x, Cur

y) by the router resolve own it co-ordinates because (X, Y), this present target is analyzed through the goal related to location router (Diverse x, Diverse y) used by the packet. The router routes the output based ahead the manufacturing of relationship. Assume (Diverse x > x) that is Cur in that full case the transfer connected with brain flit is usually to East, otherwise the most notable flit obtains a chance towards West awaiting (Diverse x, Cur x) come to be make equal which is often term because Horizontal Alignment. Whenever (Des y, Cur y) meet up through compression, with condition (Des y < Cur y) consequently be detected the header flit for any packets transfer to Southern, consider on some possibility (Diverse y=Cur y) followed by the header flit transfer to North.

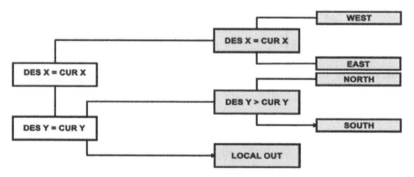

FIGURE 9.11   Flow chart of 2D routing algorithm.

Figure 9.11 illustrates the analysis that is theoretical of NoC which includes 16 nodes this basically means. 4×4 communities, the system is intended with work of Mesh topology everywhere information is communicated over the community into conditions of packet switching and routing with these particular packets is supported at XY routing method. The analysis of 2×2 operational system which includes 4 nodes is analyzed in the before topics regarding the component [9, 15].

9.8.7   FORMAT OF PACKET FORMATTING IN 2D ROUTING

The model works the 48-bit packet during the technique that is created. It is actually counted as bits on 0 towards 47. The packet arrangement is finished because exposed in Figure 9.10. First the bits on 0 and 1 is employed for location address Y, the following and bits which may be third helpful for location address of X, the bits on 4 to 15 are unused, additionally bits on 16 to 47 can be used for information out/packet out for routing, the demand bit is scheduled once upon a right time this will be completed the programer the nodes begin transfer data one considering an additional. The packets that might have request bit is techniques in XY routing algorithm, nevertheless when an entire wide range of packets are until with a duration used for pointed information regarding the memory element, on this instant the concern encoder describe the concern based ahead the select lines concerning the intermediary considered in the arrangement (Figure 9.12) [11, 15].

FIGURE 9.12   Format of parcel arranging.

The concern encoder allocates the concern within conformity towards the intermediary. Consequently to, deadlock condition is prevented through data transfer. Figure 9.13 illustrates the priority encoder intended for the method, the concern encoder has a packet period of 48-bit packet size, but consider Node 1 close to this indicate, the inputs towards this node may be on east otherwise south, now possible are there those packets may get to the indicate that is same of, therefore priority workings that are encoder accordance choose lines from the encoder. Thus, which increasingly has the greater concern, that packet is chosen ideal for extra process. Correctly, packet that has concern that is highest is allowed to ahead through the network.

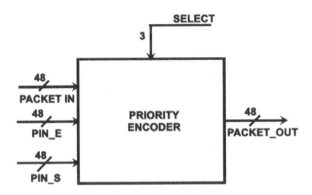

FIGURE 9.13   Priority encoder.

9.9   FAULT-TOLERANT ROUTING ALGORITHMS

Courses want that cached into a chip multiprocessor node's memory that is local [3, 4]. PE is presented after rearranging otherwise in method reconfiguration mistake discovery that is adhering to. The router operations since following to the moment that is first the target of recap energetic easy paths (optimum dimension $(N-1)$ jumps, anywhere $N = k \times k$ is entire range nodes) attaching every set of nodes:

- The trail for any kind of output port (age. g by the idea's node, send PE package towards every organized area manufacturing port with record. North, West, South, East that is otherwise the resultant package.
- By every past node, move the package in the direction of the node that is existing to get program details with ahead the package near simply visited next-door neighbors. Details course info into transfer packets and transform details into forwarded packets. Send out PE package due to the fact that a go once more packet in the direction of the foundation node problems, it is the really extremely initial packet to be right here for a supply that is specific. Put the path away in to the collection that is ahead it manages some node used for that the present node will not have a proper program, moreover eradicate the training course.
- While come back packets make it work for the supply node, from then on reserve the trail information meant for check out nodes that are seeing the course in just a method related in direction of to be able to think about into action 2. The beyond algorithm can be comprehensive additional near collection extra than one program helpful for a source-destination set that is particular. Relevant notion of the algorithm that is above combination with active execution categorization could be establish away. This inspiration regarding the router architecture that is micro performance possibility (Figure 9.14).

In the duration that is 2nd transmission that is typical in with cached courses, also the router essentially functions comparable a supply wormhole router to path variable size packages, depending in advance the paths provided in the structure node. Topologies care for near develop into unequal following errors since of this arbitrary type of mistakes, to see that is table approaches have actually currently been suggested to the previous made use of by this description, as a substitute of the state that is limited based method. This attribute that is specific specifically ideal developed for the future formula that is two-phase of paths as well as providing data utilizing cached courses.

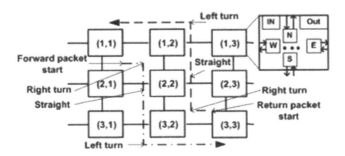

FIGURE 9.14   Relationship in between forward and return paths.

9.9.1   ROUTING AND RELIABLE THAT IS ADAPTIVE ALGORITHMS

Adaptive routing, also called active routing, is really a strategy employed for determining the perfect path a data packet must result via a network getting here on a destination that is particular [3]. Adaptive routing can be evaluated to a fare attractive a route that is different work after learning how exactly to traffic at their regular path is supported [4]. The street unavailable due to a packet could not merely depend on its supply and location apart from moreover during the present condition within the network in this partition we phrase on adaptive routing, quite simply. Adaptive routing has some benefits above insensible routing:

- It is feasible to comprehend top-quality routing outcomes among deterministic routing that is adaptive, but we recognize that the Borodin-Hopcroft lower bound to deterministic routing that is insensible can work inadequately into particular conditions.
- Various routing that is adaptive can adjust rapidly towards defective nodes otherwise boundaries towards the community, but within insensible routing defective nodes or even boundaries could detach dependable source-destination pairs with hence can require the calculation of various programming system in an exclusive manner.
- The obstructive of adaptive routing draws near are able to be there a lot better than exactly what do be gets with insensible routing approaches.

Yet, adaptive routing often furthermore has different faults:

- A freedom that is most of a path useful for a packet can reason high interruptions representing the packets, though it could get yourself a period awaiting the algorithm join to excellent paths.
- Adaptive routing could design transmission (that is, extra for the reason why too many of them need manage packets) with could contain greater load during the hardware and software than insensible routing (prize the algorithms below).
- The evaluation of flexible transmitting is regularly plenty more difficult than producing for insensible transmitting. We dedication mostly limit ourselves towards there formulas in this element.

Flexible routing on a regular basis operates at a variety that is packet-based along with packet could realize a program that is various [3]. Therefore, it is over and over probably not practical in the direction of simplifying the situation of routing details when collaborating with a flexible algorithm in the direction of a difficulty of circulation issue. As a result of the system, it is ideal for insensible routing although we did. As an option, we will take on alongside info is sent out to the style of unit-size packets. All node has been a total result called barriers right into which packages within transfer understand just how to be right here preserved. Every single time, a package to gain access to its destination it is engaged. Towards keep our models very easy basic, we approve to all or any kind of approach handle to connect merely one packet into every instruction inside a correct time action.

Some flexible transmitting strategies have currently been recommended in the direction of the past. Those inhuman pounds are:

1. **Flexible Transmitting Based at Courses Brings in Near (Figure 9.15):** Currently course approaches associated in the direction of a program that is insensible may be particular. Nonetheless, as a substitute of randomly pick selecting the training course valuable for a packet on its selections, a course can be chosen based on the condition that develops the device.
2. **Flexible Transmitting Based at Side Prices:** During these circumstances, all side is related to an expense, with the objective is frequently in the direction of trying to find a path utilized by a package with very little price. This is in fact the model technique of transmitting procedures found in to your Internet (e.g., HOLE with OSPF).

3. **Adaptive Routing Based at Neighborhood Information (Figure 9.16):** Now, no course technique benefit that is otherwise is recognized. All nodes simply make use of information that is local (for instance, packets presently living with a node) on itself alongside its neighbors that are direct choosing which packet towards communicate near to an advantage. This is certainly plainly the hardest concerning the three instances, as no support that is general instance course models or else part prices is available. We dedicate share with an algorithm used for a few of these circumstances.

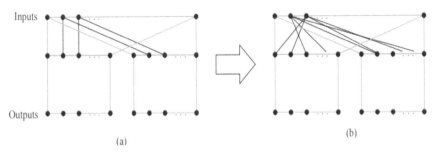

(a)

(b)

FIGURE 9.15   The image (a) illustrates a butterfly that is standard; and (b) illustrates a variation from it through which all node has some sides to each half.

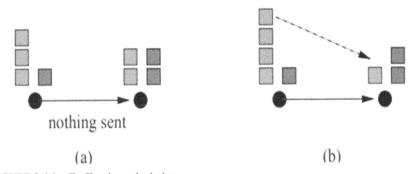

nothing sent

(a)   (b)

FIGURE 9.16   T-offsetting calculation.

9.10   CONCLUSION

The strategy by which bundles are guided from source to destination can be deterministic, unaware, or versatile. Systems with no way differences are compelled to utilize deterministic directing techniques; different topologies may utilize unaware or versatile calculations. Unaware calculations course bundles with no learning of the condition of the system. Generally, the approach of the routing is to go through the different routing algorithms for regular and irregular topologies in network on chip.

KEYWORDS

- **deadlock**
- **fault tolerance**
- **flow control**
- **mesh network**
- **quality of service**
- **routing**

REFERENCES

1. AshokKumar, N., Nagarajan, P., Sathish, K. S., & Venkatramana, P., (2020). Design challenges for 3-dimensional network-on-chip (NoC). *ICSCN 2019, LNDECT, 39*, 773–782.

2. Ashok, K. N., Nagarajan, P., Vithyalakshmi, N., & Venkataramana, P., (2019). *Quad-Rail Sense Amplifier Based NoC Router Design* (pp. 1449–1454). ICICI 2018, LNDECT 26. Springer.

3. Boraten, T., & Kodi, A. K., (2018). Runtime techniques to mitigate soft errors in network-on-chip (NoC) architectures. *IEEE Trans. Comput. -Aided Des. Integr. Circuits Syst., 37*(3), 682–695.

4. Salamat, R., Khayambashi, M., Ebrahimi, M., & Bagherzadeh, N., (2018). Design and evaluation of high-performance and fault-tolerant routing algorithms for 3D-NoCs. *IEEE Trans. Comput.*

5. Naik, A., & Ramesh, T. K., (2016). Efficient network on chip (NoC) using heterogeneous circuit-switched routers. In: *International Conference on VLSI Systems, Architectures, Technology and Applications (VLSI-SATA)*.

6. Ashokkumar, N., & Kavitha, A., (2016). A novel 3D NoC scheme for high throughput unicast and multicast routing protocols. *Tech. Gazette, 23*(1), 215–219.

7. Ashokkumar, N., & Kavitha, (2015). Network on chip: A framework for routing in system on chip. *Journal of Computational and Theoretical Nano Science, 12*(12), 6077–6083.

8. Nallathambi, G., & Rajaram, S., (2014). A particle swarm optimization approach for low power very large-scale integration routing. *J. Math. Stat., 10*, 58.

9. Viswanathan, N., Paramasivam, K., & Somasundaram, K., (2012). Exploring optimal topology and routing algorithm for 3D network on chip. *Am. J. Appl. Sci., 9,* 300–308.

10. Michelogiannakis, G., Becker, D., & Dally, W., (2011). Evaluating elastic buffer and wormhole flow control. *IEEE Trans. Computers, 60*(6), 896–903.

11. Murugappa, P., Al-Khayat, R., Baghdadi, A., & Jezequel, M., (2011). A flexible high throughput multi-ASIP architecture for LDPC and turbo decoding. In: *Proceedings of the Design, Automation and Test in Europe Conference and Exhibition (TECE)* (pp. 1–6).

12. Kao, Y. H., & Chao, C. J., (2011). *5th Conference of IEEE/ACM International Symposium on Networks on Chip, 81.*

13. Michelogiannakis, G., Sanchez, D., Dally, W. J., & Kozyrakis, C., (2010). Evaluating bufferless flow control for on-chip networks. In: *Proceedings of IEEE/ACM Fourth International Symposium on Networks-on-Chip (NOCS'10)* (pp. 9–16).

14. Ramanujam, R. S., Soteriou, V., Lin, B., & Peh, L. S., (2012). Design of a high-throughput distributed shared buffer NoC router. In: *Networks-on-Chip (NOCS).*

15. Zhang, W., Hou, L., Wang, J., Geng, S., & Wu, W., (2009). Comparison research between XY and odd-even routing algorithm of a 2-dimension 3X3 mesh topology network-on-chip. In: *WRI Global Congress on Intelligent Systems, GCIS 2009* (Vol. 3, pp. 329–333).

CHAPTER 10

Self-Driven Clock Gating Technique for Dynamic Power Reduction of High-Speed Complex Systems

ROOPA R. KULKARNI[1] and S. Y. KULKARNI[2]

[1]*Department of Electronics and Communication Engineering, Dayananda Sagar Academy of Technology and Management, Bengaluru, Karnataka, India, E-mail: roopa.patavardhan@gmail.com*

[2]*REVA Institute of Technology and Management, Bengaluru, Karnataka, India*

ABSTRACT

Clock gating is one of the predominant technologies used for saving power in particular, the dynamic power. It is observed that the existing clock gating such as AND based and latch-based techniques do not have the concept of feedback. The concept of feedback through a comparator adds the advantage of reducing the power of the system. In this chapter, a self-driven clock gating technique is proposed that uses the XNOR gate as the comparator. The technique compares the present input to the previous output and then finally the output is ANDed to obtain the gated output. This technique can be applied to any high-speed systems such as: DSP processors, image processor, bio-medical processors and in general to any processor that requires power optimization. The authors of this chapter have applied this technique to an academic state-of-art design of 32-bit RISC processor, achieving a total power reduction of nearly 16% to 18%, with an increase in the area of 3% to 5% for different set of test cases, using 45 nm process technology. This can be further integrated into an Electronic Design Automation commercial backend design flow, wherein the algorithm of the tool can select the proposed clock gating techniques as the integrated clock gating cell.

10.1   INTRODUCTION

Computing technology is a major catalyst for innovation among various technologies, which affects every aspect of modern society. Semiconductor technology along with computer architecture provides the necessary infrastructure for the development of high performance computationally intensive application with low-energy operations. The tremendous growth in the embedded application, which are battery operated and raw power-sensitive application, have created a global demand for low power semiconductor industry.

In addition, with the growth in the technology of wireless communication, multimedia, server, router, automotive applications as well as and medical, the impact of low power design has become a bottleneck. Low power techniques, when used, could also increase the product development time. The factors that contribute to the increased complexity, throughout the design flow are; additional functional verification at RTL, complexity during synthesis and layout, increased need for special library characterization and area increase due to additional logic needed to support low power design.

Hence, there is a need to reduce the power of the systems that are used in various real time applications. Clock signal is the highest frequency toggling signal in a system-on-chip (SoC). The toggling rate of the applied frequency is a proportional impact of the capacitive load power component of the dynamic power of a device. Hence, the buffers and the inverters in the clock path, would contribute highest to the dynamic power consumption [1] in SoC. Power consumption in the clock path alone contribute to more than 50% of the total dynamic power consumed within modern SoC. Power being a very significant aspect of the design, efforts need to be made to reduce this. Clock gating [7] is one such method.

The clock gating logic works on the principle of gating the nonfunctional logical unit or nonfunctional high-performance latches from the clock input. This technique can be applied at all levels: system level, architectural design, logic as well as gate-level [2, 3]. The techniques to reduce the dynamic power using clock gating along with the advantages are described in Refs. [4, 6]. The clock signal can be either gated to a latch or a logical gate by ANDing the clock input with a control signal. Hence, preventing the redundant charging/discharging of the capacitances when the circuit is idle. This helps in saving the circuit's clock power. Fundamentally, there are two basic types of clock gating, latch free and latch-based clock gating. The latch free clock gating technique uses an AND based logic, wherein the input to the AND gate is the enable signal and the clock input. The clock is gated when the enable input is low and is transparent when the enable input is high. The drawback of this is that it leads to the generation of glitches, which leads to uncertainty in the gated clock output. To overcome this major drawback a latch-based clock gating is used. In this technique, the gated clock output is obtained from the latch and the AND gate. The latch will hold the enable input till the clock input and hence overcomes the glitch problem. The other flavor of the architectural clock gating techniques are: data-driven CG, look ahead CG, self XORed CG, etc.

The major concern of the existing clock gating techniques is; the clock enabling signals is applied to a group of FFs that are highly correlated; physical locations of FFs affect the delay and clock skew; desirable for FFs driven collectively, by the same clock gate, to be placed close to each other. Data-driven gating suffers from a very small time window during which the gating circuitry can work properly. The total delay of the XOR, OR, latch, and the AND gate should not exceed the setup time of the flip flops. This is the motivation to address these drawbacks in the self-driven clock gating. This chapter presents a self-driven clock gating technique, its working principle, the methodology and finally the technique is applied to a 32-bit RISC processor to justify the performance parameters.

The chapter is organized as: Section 10.2 discusses on the related work carried out in the area of clock gating technique; Section 10.3 discusses the proposed clock gating technique; Sections 10.4 and 10.5 provide the detailed description of the gating methodology and implementation. Finally, to justify the technique discussed, a target 32-bit RISC processor is considered for analyzes. Sections 10.6 and 10.7 describe the application self-driven clock gating technique to a 32-bit RISC processor to compare and analyze the performance parameters namely: area and power. The concluding remarks on the work carried out are presented in Sections 10.7 and 10.8.

10.2   RELATED WORK

Several methods and variations are built around the fundamental clock gating technique that are surveyed and described. Sequential circuits are the major building blocks of any digital system. The operation of these circuits depends on the clock applied, which can be the master clock or the clock derived from the master clock circuitry. Recent studies have shown that the clock circuitry consumes about 15%–45% of total power. Hence, the need of studying the behavior of the clock and hence providing the clock gating is described in Ref. [4]. The authors study the behavior of the clock at each transistor and arrive at two important terms namely: transition propagate and generate. The authors also describe the design of sequential circuits through examples based on derived clock. Thus, to conclude derive clock could be used to isolate the triggered flip-flop from the master clock during its idle cycle.

One of the predominantly used technique to reduce the unwanted switching of the clock input signal is clock gating. This technique can be employed at all levels of the design namely: system, architecture block, and logic design. In Ref. [8], the authors describe the existing clock gating techniques that is, clock gated positive latch, data driven clock gating and auto gated flip-flops. Each of these have the drawback of either the flip-flop overhead, or short time window or the timing constraint imposed. To overcome, these drawbacks a look ahead clock gating technique is proposed which computes the clock enabling signal of each flip-flop, one cycle ahead of time, depending on the present input cycle. This technique is applied to 4, 8, 16, 32, and 64-bit linear feedback shift register (LFSR). LFSR is the well-known and most commonly used pseudo-random number generator circuit used in the built-in self-test, signature analysis and in spread spectrum communication. The results have shown, an average of power saved is in the range of 27.8% to 42.21% depending on the n-bit of the LFSR.

In Ref. [9], the authors give an overview of intelligent clock gating for power optimization employed in Xilinx FPGA. Xilinx has the automated capability linked to place and route, which uses a set of innovative algorithms to analyze the design. To neutralize the superfluous switching activity, the analysis and detection of the source register contributing to the result for each clock cycle is obtained. The software then adds the abundant amount of clock enable to the logic to create fine-grain clock gating that neutralizes superfluous switching activity. The clock enables added to optimize the design do not alter the pre-existing logic or clock placement, nor do they clear new clocks. There is an additional logic of 2% of LUTs is created. Additional optimization is used to reduce the power for dedicated block RAM using intelligent clock gating. Thus, to conclude, the chapter deals with the Xilinx tool which provides the clock gating to the entire design without any changes to existing logic or clock that would alter the behavior or timing of the original design.

In Ref. [10], the author proposed the technique of "XOR self-gating" that can gate the clock during the absence of enable and reset signals. The technique is based on the observation that if a register output and input are the same, then no new data is required to be loaded and hence, no clock signal is required to be applied. This technique is implemented based on two criteria that are efficiency and locality. From the study and implementation, it is observed that XOR self-gating insertion can be performed using RTL or at gate-level netlist, also compatible with usage of unified power format (UPF). To conclude, in order to meet the power budget of complex designs, power management has become a challenging task. Hence, implementation of the clock gating technique in design compiler or IC compiler is become the necessity. Clock gating, the commonly used technique at synthesis leaves a large amount of redundant clock pulses. Data-driven gating is one of the methods that disable these. The authors in Ref. [11] aim at the grouping of FFs in such a way that maximizes the power reduction. The chapter introduces to the study of data-driven clock gating. During the operation, whenever the state of the FF is not subjected to change in the next clock cycle, the clock driving such FFs is disabled, meaning the clock is gated. Hence, to reduce the overhead due to clocking, several FFs are grouped and clocked by the same clock signal. Another method of grouping of FFs called multi-bit FF (MBFF) is also discussed. This method physically merges FFs in to a single cell, such that the master and the slave latches share the common clock pulse through an inverter.

10.3   SELF-DRIVEN CLOCK GATING

The major concern in the data driven clock gating technique is the increase in area and power of the target circuit due to overheads of additional XOR and OR gates. To overcome this drawback, the sequential logic are grouped and ORed together that can be driven by the same clock gating signal. The challenge in this technique is to identify the flip flops with similar toggling rate and their physical positions, such that they can be grouped and triggered by the same clock gating signal. The contribution of the self-driven clock gating technique is it overcomes the drawback of the small window for triggering the gated clock logic and to embed the complete clock gating logic as a standard cell into the foundry.

The self-driven clock gating technique is discussed as: clock gating logic is implemented with a Boolean logic and the latch-based CG. The Boolean

expression creates the required functional logic that gates a few idle functional blocks keeping the functionality of the target design intact. This work
is an effort made to design a functional clock gating logic for high speed
and complex circuits. The proposed clock gating technique consists of a
latch and a AND gate that avoids the generation of glitches into the circuit.
Another feature added to this is a comparator that compares the present D
(enable) input and the previous Q output, this data is ORed with clock input
and given as the clock input to the latch. Here, the comparator can be a
XOR, XNOR or a MUX depending on the latch being the positive edge or
negative edge. The motivation to use this logic is that it is self-driven, which
reduces the power by an additional latch power. Firstly, clock gating logic is
implemented as a Boolean function that logically added the required delay
and functionally the target circuit does not fail. Secondly, the comparator can
be a part of the complete clock gating standard cell. This makes the complete
design compact, which overcomes the drawback of the small window for the
latch to trigger the further logic. The logic diagram of the self-driven clock
gating circuit is as shown in Figure 10.1 and the relevant waveforms are
shown in Figure 10.2.

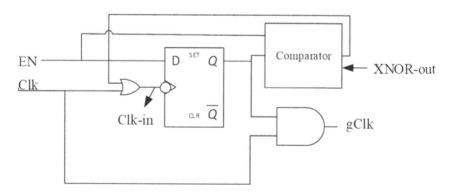

FIGURE 10.1   Self-driven low power clock gating.

A gated clock signal is generated by the clock gating cell based on
an input clock signal and the output signal. The generated clock signal is
provided to a clock input stage of the latch circuit in the clock gating cell. The
switch in the input signal, switches the latch clock signal. This eliminates the
continuous switching of the input signal at the clock input stage of the latch
circuit. Hence, this reduces the charging and discharging of a capacitive load

associated with the clock input stage of the latch circuit. Thus, reduces the power loss of the clock gating cell due to switching activity.

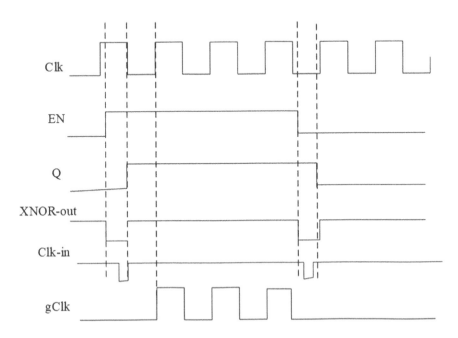

FIGURE 10.2   Waveform at every node of self-driven CG.

The clock and the enable input are provided to the gating circuit. The comparator is the XNOR gate that compares the present input EN given as the D input with the previous output Q. If the output of the XNOR is low, and hence the low signal is provided at the gated output. On the other hand, when the input is the same as the previous output, the clk-in is high, and therefore the output of the OR gate is the clk input to the latch. This triggers the latch; the Q output is ANDed with the Clk input and a gated clock is obtained at the output. The XNOR as a comparator can be replaced with XOR or MUX circuit. The negative latch has the advantage of a zero cycle if the setup time and hold time can be met in the immediate cycle.

The implementation of the clock gating logic is done at the register transfer level (RTL) for 45 nm technology using Cadence EDA tool. The Verilog code is written for the self-driven clock gating logic, and is simulated to check for its desired functionality. The functional verification of the circuits is as shown in Figure 10.3.

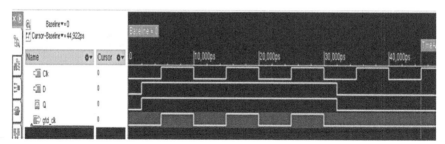

FIGURE 10.3   Simulated output of the self-driven clock gating logic.

The code is synthesized and the gate-level netlist is obtained and is further taken to the physical design to generate the layout. The physical layout is then checked for optimization with respect to design as well as timing. The negative slack, if obtained, the clock tree synthesis is performed for the setup and hold time. The RC extraction is performed to extract the parasitic capacitance and resistance. The power report of the post-synthesis is obtained. The layout is as shown in Figure 10.4.

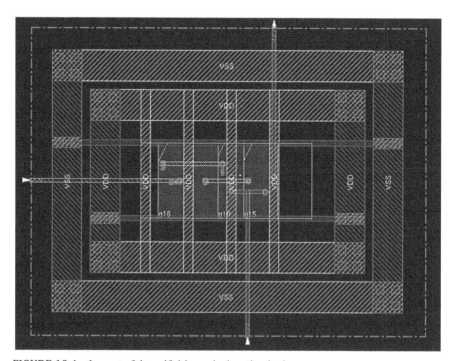

FIGURE 10.4   Layout of the self-driven clock gating logic.

The parameters of power, area, and timing are obtained for a set of test cases, and presented in Table 10.1.

TABLE 10.1   Post Synthesis Parameters

Parameters	Value
Power	0.00031867 W
Area	1.0260 μm²
Timing	1.943 ns

10.4   METHODOLOGY

The methodology and the process of implementing the self-driven clock gating as part of the complete system design flow describes the working principle of the self-driver logic and hence its effect at the system-level design. The process involved in the implementation of the clock gating to generate the required gated clock is given in Figure 10.5.

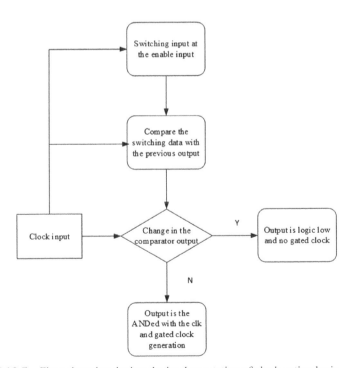

FIGURE 10.5   Flow chart that depicts the implementation of clock gating logic.

The clock input is provided to every stage of the design flow. The output signal change is based on both the latch clock signal and the input signal. The clock gating cell consist of three components, namely: a comparator logic circuit, a first logic circuit, and a second logic circuit. The comparator circuit is connected to at least one of the output stages. The input and the output signals are compared by the comparator to generate the required output that decides the clocking output. The input to the first logic circuit is provided from the output signal. The clock input stage gets the output from the first logic circuit and switches an input clock signal between a first state and a second state, hence, generating the latch clock signal. Finally, the second logic circuit receives inputs from clock signal and the output signal, that generates the required gated clock signal.

The clock gating cell includes a latch circuit, which receives the input signal and a clock signal. An output signal obtained is based on the input signal and the latch clock signal. Thereafter, the comparison signal is generated by comparing the input and the output signals. The input clock signal is switched between a first state and a second state based on the comparison signal to generate the latch clock signal. Finally, a gated clock signal is generated based on the input clock and the output signal.

Initially, the toggling probabilities of the system are estimated with extensive test cases. This determines the functional modules of the system to be clock gated. The clock gating circuit in particular checks previous output to that of the present input, if there is change in the data the latch is triggered on the negative edge. This is further ANDed with the enable input to generate the required gated clock output.

10.5 IMPLEMENTATION OF CLOCK GATING ON 32-BIT RISC PROCESSOR

The architecture of the 32-bit MIPS based RISC processor is designed with 5-pipeline stages. The unique feature of this architecture is it has the hazard detection unit as the part of the design. The detailed architecture consisting of the five stages namely; instruction fetch (IF), instruction decoder (ID), execute, memory access (MEM), and write back (WB) stage along with the hazard unit is as shown in Figure 10.6. The brief description of all the stage is:

1. **Instruction Fetch (IF) Stage:** This stage of the processor fetches the instruction to be executed from the instruction memory (IM). The instruction to be fetched is addressed by the program counter. The

encoding size of the instruction is 32-bit and hence the 32-bit address is held in the PC.

2. **Instruction Decode (ID) Stage:** This is the second stage of the processor, which decodes the instruction. This stage after decoding separates the operational code and operand from the instruction, to understand the operation to be performed as well as the operands to fetch the data. This stage also accesses the register file to read the register.

3. **Execute (EX) Stage:** The execution stage is the main block of the processor, where the actual computation takes place. The execute stage requires ALU to performs all the arithmetic, logical, and shift related operation based on the decoded oper-code value.

4. **Memory Access (MEM) Stage:** The RISC processors have the main feature of load/store architecture. This means that most of the data is stored in the memory called the data memory. Hence, MEM is the stage required for all the load/store operations.

5. **Write Back (WB) Stage:** Once the input data is processed, the result is stored either in the register or the memory. Hence, this stage is required to store the result into the register.

6. **Hazard Detection Unit:** The RISC processor designed has 5 pipeline stages, which implies the fetching of next consecutive instructions well in advance. In the due course of time, few instructions would depend on data, or control operations. This will halt the processor operations. To address this, the 32-bit RISC processor has hazard detection unit to detect the control hazard and data hazard. It detects the hazard and tries to reduce it by introducing stall in order to get the correct result.

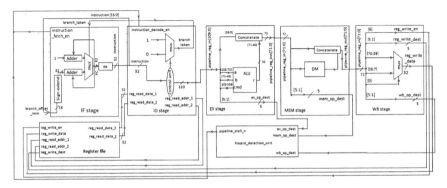

FIGURE 10.6   Architecture of the 32-bit MIPS-based RISC processor.
Source: Reprinted with permission from Ref. [26]. © 2019 Springer.

The instruction formats of MIPS considered in the design are of two types; the register type and immediate type (I-type):

1. Register Type (R-Type): The instructions considered are the 3-address format. In this, the first operand is the destination while the other two are the sources. In Register Type (R-type) instruction, the sources and the destination operands are register on which computation is carried out. Here, Src1 and Src2 are the two-source register which hold the data to be processed. The destination register, Dest is the used for storing the computed result. Table 10.2 shows the format for the R-Type instruction.

TABLE 10.2   Register Type (R-Type)

Oper-Code	Dest	Src1	Src2	Reserved
31–26	25–21	20–16	15–11	10–0

In R-Type instruction, bits 31–26 are used for Oper-code, bits 25–21 are used to indicate address of destination register (Dest), bits 20–16 and 15–11 corresponds to address of source-1 and source-2 (Src2) register, respectively, while the remaining 11-bits are reserved.

2. Immediate Type (I-Type): Among the two source operands, one of the operands can be an immediate data, while the other one can either be the register or memory. The branch instructions have label as the immediate data referenced through the name. The immediate value in case of branch instruction will act as the offset which is added to the PC to obtain the effective address. Thus, all such instructions can be categorized as I-type. Table 10.3 shows the format for the I-Type instruction.

TABLE 10.3   Immediate Type (I-Type)

Oper-Code	Dest	Src1	Immediate Value
31–26	25–21	20–16	15–0

In I-type instruction bits 31–26 corresponds to Oper-code, bits 25–21 corresponds to the address of the destination register (Dest), bits 20–16 are used to represent the address of the source-1 register (Src1). The remaining 16-bits are used as an immediate value. In case of branch operation, it is considered as offset address.

The various instructions supported by the 32-bit RISC processor are tabulated in Table 10.4. Table 10.4 describes the various operations supported by the architecture, the corresponding operational code and the specific operation. The instructions are written in accordance to the instruction format that is either I-type or R-type. The set of instructions are embedded into the register file from where the program counter register will access instruction by instruction.

TABLE 10.4   MIPS Instructions

SL. No.	Opcode	Operation	Description
1.	000000	OPER_NOP	No operation
2.	000001	OPER_ADD	Res= A+B
3.	000010	OPER_SUB	Res= A-B
4.	000011	OPER_AND	Res= A&B
5.	000100	OPER_OR	Res= A|B
6.	000101	OPER_XOR	Res= A^B
7.	000110	OPER_NOT	Res= ~B
8.	000111	OPER_SL	Res= A<<B
9.	001000	OPER_SR	Res= A>>B (sign ext)
10.	001001	OPER_SRU	Res= A>>B
11.	001010	OPER_LD	Rd= mem[Rs]
12.	001011	OPER_ST	mem[Rs]=Rt
13.	001100	OPER_BZ	Branch pc=pc+offset
14.	100001	OPER_ADDi	Res= A+imm
15.	100010	OPER_SUBi	Res= A-imm
16.	100011	OPER_ANDi	Res= A&imm
17.	100100	OPER_ORi	Res= A|imm
18.	100101	OPER_XORi	Res= A^imm
19.	100110	OPER_NOTi	Res= ~imm
20.	001101	MOV	Res= B
21.	101101	MOVi	Res=imm

The data flow that happens in the RISC processor is shown in Figure 10.7. All the instructions are stored in the IM. The program counter register holds the address of the next instruction to be executed. The IF fetches the instruction from the IM. This instruction is decoded by the decode stage that separates the operational code and the operands. At this stage, the processor knows what operation is to be performed. The execute stage performs the computation of the required operation with the help of the ALU. MEM stage loads or stores the result. Finally, WB stage stores the result back into the destination register.

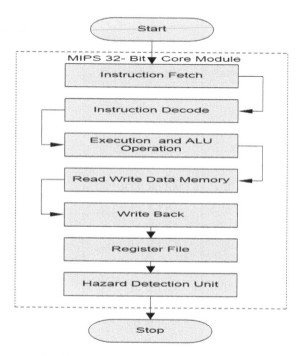

FIGURE 10.7   Flowchart for 32-bit RISC processor.
Source: Adapted with permission from Ref. [26]. © 2019 Springer.

10.6   SIMULATION RESULTS

To begin with, the designed RISC processor is simulated for functional verification, the code is then synthesized to obtain the gate-level netlist. The power, area, and timing reports are obtained after synthesis. The RISC processor is simulated with a set of instructions that cover the register as well as immediate data type with an overall coverage of 75%. The clock input is provided to all the stages of the processor. The stall signal occurs whenever there is a need for the resource, and as seen from the simulated result, the corresponding register file as well as the IF hold the same data but, the clock signal is provided through no operation is performed. The clock signal makes transitions continuously when no operation is performed. This indicates that the register file as well as the IF can be put to sleep mode whenever the stall signal occurs.

The next stage is to incorporate the proposed clock gating technique into the RISC processor identifying the functional block that makes maximum transition, and the functional block that is idle for a specified duration of time or for a specific set of operations. This is a very challenging task at the RTL level, as the target code is written with the behavioral description and identifying the sequential logic that need to be gated is a difficult to figure out. To overcome and meet this challenge to a larger extend, the RISC processor is designed with a top-down approach, wherein the modularity is maintained and consists of sub-modules. This aids in identifying the functional units that can be gated when idle.

Out of the five functional blocks, only three blocks in architecture have the control signal namely; "IF, ID and register file." Hence the self-gated clock gating technique is applied to register file and to the IF while not to the ID stage, it will be stalled automatically as it comes immediately after the IF stage in case of hazard. The clock gating cell blocks the clock to these two blocks when they go idle. The stall signal is the control signal given to the clock gating circuit as it determines the gating operation. The output of this logic provides a glitch free clock gating to the logic. The architecture of 32-bit RISC processor is simulated with and without the application of clock gating logic and for the same set of instructions.

10.7   IMPLEMENTATION

The functional verification of the proposed clock gating technique is then synthesized to obtain the required report of the performance parameters namely area, power, and delay. The 32-bit RISC processor with and without the clock gating technique is implemented for a 45 nm technology library using Cadence EDA tool. The constrains for the clock, define the frequency of operation, the rise time, fall time, minimum drive strength, and the effort of optimization. The clock frequency of 250 MHz is considered for the implementation and evaluation. On synthesis, the gate-level netlist is generated. The gate-level netlist defines the standard cells used in the implementation of the RISC processor. The choice of the standard cells for the target code depends on the operating frequency, depth, and functionality of logic to be implemented, as well as the load capacitance. The schematic view after synthesis shows the internal logic implemented as well as the complexity of the design. The complete design with the insertion of self-driven clock gating logic is as shown in Figure 10.8. The performance of the RISC processor is evaluated for power, area, and timing. The behavior of the design and its performance are reflected by the input test cases provided to the logic. The design is evaluated with respect to the following framework:

- Initially, simulate, and synthesize the target RTL design without the application of clock gating, providing the actual test cases as a set of instructions to the processor stored in the form of VCD file, and obtain the performance parameter.
- Then, simulate, and synthesize the target RTL design with the application of self-driven clock gating, with the same set of test cases VCD file, and obtain the performance parameter.
- Compare the two results for power, area, and timing.

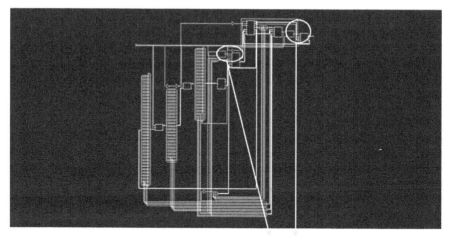

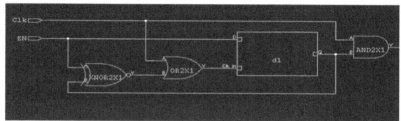

FIGURE 10.8 Top module of RISC processor with proposed clock gated logic.

10.8 RESULTS AND DISCUSSIONS

The area, power, and the timing report of a 32-bit RISC processor is acquired after the synthesis. Tables 10.5 and 10.6 represent the power and area of a 32-bit RISC processor without and with clock gating logic along with the actual test cases. The actual switching activity depends on the test cases provided at the time of simulation to generate the VCD file. In this design case, the test case is the different instructions that are executed by the program counter that in turn depends on the set of instructions stored in the IM. The test cases considered is set of instructions with an overall coverage of 75%. The instances column presents the modules defined in the processor design as well as the instances which are the standard cells selected by the tool algorithm to meet the design specifications.

The self-driven clock gating technique is now applied to the RISC processor at two stages of the pipeline. This implies there is a need of two clock gating cells that can generate two clock gated signals. The control logic used to the enable input of the CG is critical and important; thus, in this case study of 32-bit RISC processor, the stall signal is identified as the input control signal to the enable of the CG. From the simulation results, it is proved that the required gated clock signal is generated and the circuit works logically and functionally. The power reports are obtained on the similar lines, with the application of the actual switching activity and the values are tabulated in Tables 10.5 and 10.6.

TABLE 10.5 Area and Power of 32-Bit RISC Processor Without Clock Gating Logic

Instance	Area (μm²)		Power (mW)		
	Cells	Cell Area	Leakage Power (mW)	Dynamic Power (mW)	Total Power (mW)
mips_32_core_top2	46293	157996	1.099	52.098	53.197
MEM_stage_inst	28472	101817	0.746	40.275	41.021
Dmem	28257	100741	0.736	39.958	40.694
register_file_inst	7267	26662	0.195	9.376	9.571
IF_stage_inst	8050	21380	0.105	0.436	0.541
Mem	7597	19870	0.095	0.296	0.391
add_20_15	224	634	0.004	0.003	0.006
inc_add_22_15_1	31	145	0.001	0.003	0.005
EX_stage_inst	1770	5236	0.03	0.277	0.307

TABLE 10.5 *(Continued)*

Instance	Area (µm²)		Power (mW)		
	Cells	Cell Area	Leakage Power (mW)	Dynamic Power (mW)	Total Power (mW)
alu_inst1	1549	4144	0.02	0.013	0.033
add_29_18	263	716	0.004	0.001	0.004
sub_30_18	270	713	0.004	0.006	0.01
srl_37_19	237	666	0.003	0.001	0.004
sll_35_20	239	665	0.003	0.001	0.004
ID_stage_inst	548	2425	0.021	0.672	0.693
WB_stage_inst	97	242	0.002	0.001	0.003
hazard_det. ion_unit_inst	89	234	0.001	0.002	0.003

TABLE 10.6   Area and Power of Clock Gated RISC Processor with Clock Gating Logic

Instance	Cells	Cell Area	Leakage Power (mW)	Dynamic Power (mW)	Total Power (mW)
mips_32_core_top3	46636	158208	1.099	43.3	44.399
MEM_stage_inst	28507	101899	0.746	40.605	41.351
Dmem	28293	100825	0.736	40.288	41.024
IF_stage_inst_mem	7431	19489	0.095	0.285	0.381
EX_stage_inst	1305	4117	0.025	0.278	0.302
alu_inst1	1089	3037	0.014	0.014	0.028
srl_37_19	259	699	0.003	0.001	0.005
sll_35_20	249	682	0.003	0.002	0.005
ID_stage_inst	608	2516	0.021	0.534	0.554
selfcg_inst	6	24	0	0.013	0.013
selfcg1_inst	5	20	0	0.008	0.008

The power comparison between the non-gated and the gated RISC processor is as shown in Figure 10.9. The overall power of the processor reduces by 16.34% with an increase in the area of 0.13% as seen from the graph shown in Figure 10.9. The power reduction takes place due to the stall signal connected to the enable of self-driven clock gated logic, which gates the register file whenever this signal is one. Hence, the power of the register file reduces as there is no clock input, which implies no transition of the clock signal.

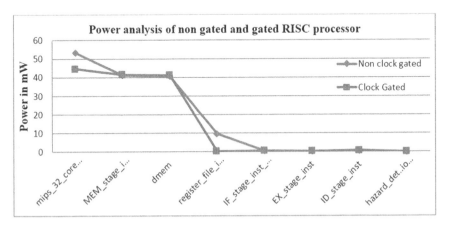

FIGURE 10.9   Area comparison of non-clock gated and clock gated RISC processor.

On comparison of instances types, it is observed that area of the logic of clock gated RISC is increased as compared to the area of non-gated RISC processor and is shown in Table 10.7. It is observed that the sequential logic instances have increased along with the logic and the buffers, due to the addition of self-driven clock gating logic while, the number of inverters has reduced.

TABLE 10.7   Area Comparison of Non-Clock Gated (CG) and Clock Gated RISC Processor in Terms of Type of Instances

Type	No. of Instances		Area (μm²)	
	No CG	CG	No CG	CG
Sequential	5,470	5,472	56473.22	56491.52
Inverter	2,297	1,071	3263.043	1536.019
Buffer	17,274	18,367	40533.44	43102.39
Logic	21,252	21,726	57726.25	57077.67

10.9 CONCLUSION

This chapter presents a self-driven clock gating technique that works on the feedback through a comparator. The negative latch is used hence, lead to zero cycle. This implies that the setup and hold time can be set on the immediate edge of the clock cycle. On the comparator output being one, the latch will not transfer the enable and hence the gated clock output will not be triggered. This leads to an additional power saving. To prove the efficacy of the technique, it is applied to a 32-bit RISC processor. Results have shown a total power reduction of 16.34% with the code and functional coverage of 78% to 80%. There is a small increase in the area of 3% to 5% and meets the performance requirement. Presently, this technique is manually inserted at the RTL and it can be further taken into the EDA backend flow.

KEYWORDS

- **clock gating**
- **data-driven clock gating**
- **dynamic power reduction**
- **instruction memory**
- **linear feedback shift register**
- **RISC processor**

REFERENCES

1. Oklobdzija, V. G., (2000). *Digital System Clocking – High Performance and Low Power Aspects.* New York, USA: Wiley,
2. Benini, L., Bogliolo, A., & De Micheli, G., (2008). A survey on design techniques for system - level dynamic power management. *IEEE Trans. Very Large Scale Integr. (VLSI) Syst., 8*(299–316).
3. Hosny, M. S., & Yuejian, W., (2002). Low power clocking strategies in deep submicron technologies. *Proc. IEEE Intll. Conf. Integr. Circuit Design Technol.,* 143–146.
4. Chunhong, C., Changjun, K., & Majid, S., (2001). Activity-sensitive clock tree construction for low power. *Proc. Int. Symp. Low Power Electron. Design,* 279–282.
5. Farrahi, A., Chen, C., Srivastava, A., Tellez, G., & Sarrafzadeh, M., (2001). Activity-driven clock design. *IEEE Trans. Comput. Aided Design Integr. Circuits Syst., 20*(6), 705–714.
6. Shen, W., Cai, Y., Hong, X., & Hu, J., (2008). Activity and register placement aware gated clock network design. In: *Proc. Int. Symp. Phys. Design* (pp. 182–189).

7. Oliver, J. P., Curto, J., Bouvier, D., Ramos, M., & Boemo, E., (2012). Clock gating and clock enable for FPGA power reduction. In: *8th Southern Conference on Programmable Logic (SPL)* (pp. 1–5).

8. Manjith, R., & Muthukumari, C., (2015). Dynamic power reduction in sequential circuits using look ahead clock gating technique. *International Journal of Electronics and Communication Engineering, International Scholarly and Scientific Research & Innovation, World Academy of Science, Engineering and Technology, 9*(2), 252–258.

9. Rivoallon, F., (2011). *Reducing Switching Power with Intelligent Clock Gating.* White Paper: Virtex-6, Spartan-6, Kintex-7, and Virtex-7 FPGAs, WP370 (v1.3).

10. Ezroni, (2011). *Advanced Dynamic Power Reduction Techniques: XOR Self-Gating.* White paper.

11. Wimer, S., & Koren, I., (2014). Design flow for flip-flop grouping in data-driven clock gating. *IEEE Transactions on Very Large-Scale Integration (VLSI) Systems, 22*(4), 771–778.

12. Neenu, J., Sabarinath, S., & Sankarapan, D. K., (2009). FPGA based implementation of high-performance architectural level low power 32-bit RISC Core. *International Conference on Advances in Recent Technologies in Communication and Computing, Annual IEEE,* 53–57.

13. Shofiqul, I., Debanjan, C., Manoja, K. D., Neelima, V., & Rahul, S., (2006). Design of high-speed pipelined execution unit of 32-bit RISC processor. *India Conference, Annual IEEE,* 1–5.

14. Ji-Hoon, K., Duk-Hyun, Y., Ki-Seok, K., Eun- Joo, B., WonHee, S., & In-Cheol, P., (2008). Design of high performance 32-bit embedded processor. *International SoC Design Conference.*

15. Soumya, M., & Usha, V., (2015). FPGA based implementation of power optimization of 32-bit RISC core using DLX architecture. *International Conference on Computing Communication Control and Automation.* IEEE.

16. Aneesh, R., Vinayak, B. P., David, S., & Vivian, D., (2016). A RISC-V instruction set processor-microarchitecture design and analysis. *International Conference on VLSI Systems, Architectures, Technology and Applications (VLSI-SATA).* IEEE.

17. Priyanka, T., & Rajan, P. T., (2015). Design and analysis of 16-bit RISC processor using low power pipelining. *International Conference on Computing, Communication and Automation (ICCCA2015).* IEEE.

18. Bharadwaja, P. V. S. R., Ravi, T. K., Naresh, B. M., & Neelima, K., (2015). Advanced low power RISC processor design using MIPS instruction set. *IEEE Sponsored 2nd International Conference On Electronics And Communication Systems(ICECS '2015).* IEEE.

19. Narender, K., & Munish, R., (2015). Implementation of embedded RISC processor with dynamic power management for low-power embedded system on SOC. *Proceedings of 2015 RAECS.* UIET Panjab University Chandigarh, IEEE.

20. Divya, & Deepty, (2015). A Verilog-based design of 16–bit RISC processor. *International Journal of Engineering and Technical Research (IJETR), 3*(6). ISSN: 2321-0869.

21. Imyong, L., Dongwook, L., & Kiyoung, C., (2008). ODALRISC: A small, low power, and configurable 32-bit RISC processor. *International SoC Design Conference,* IEEE.

22. Shmuel, W., & Israel, K., (2011). The optimal fan-out of clock network for power minimization by adaptive gating. *Very Large-Scale Integration (VLSI) Systems, IEEE Transactions, 99,* 1–9.

23. Hao, X., Ranga, V., & Wen-Ben, J., (2011). Dynamic characteristics of power gating during mode transition. *Very Large-Scale Integration (VLSI) Systems, IEEE Transactions, 19*(2), 237–249.

24. Jatin, N. M., Al-Hashimi, B. M., David, F., & Stephen, H., (2011). Sub-clock power-gating technique for minimizing leakage power during active mode. *Design, Automation & Test in Europe Conference & Exhibition*, 1–6.

25. Li, K. C., & Haiqing, N., (2011). Effective algorithm for integrating clock gating and power gating to reduce dynamic and active leakage power simultaneously. *Quality Electronic Design (ISQED), 2011 12th International Symposium*, 1–6.

26. Mangalwedhe S., Kulkarni R., Kulkarni S.Y. (2019) Low Power Implementation of 32-Bit RISC Processor with Pipelining. In: Nath V., Mandal J. (eds) Proceeding of the Second International Conference on Microelectronics, Computing & Communication Systems (MCCS 2017). Lecture Notes in Electrical Engineering, vol 476. Springer, Singapore. https://doi.org/10.1007/978-981-10-8234-4_27

CHAPTER 11

Optimization of SOC Sub-Circuits Using Mathematical Modeling

MAGNANIL GOSWAMI

Accendere Knowledge Management Services Pvt. Ltd.,
New Delhi – 110044, India, E-mail: magnanil.goswami@gmail.com

ABSTRACT

A system-on-chip (SOC) is an integrated circuit (IC) that integrates all components (analog and/or digital and/or mixed-signal and/or radiofrequency) of an electronic system in the form of a single chip. Typical low-power applications of SOC comprise cell phones, biomedical devices, consumer electronic appliances, etc. Aggressive down-scaling of semiconductors by IC manufacturers makes way for SOCs with higher packaging density. Unlike the digital integrant, technology scaling has largely affected the performance of analog and radio frequency (RF) components of low-power SOCs. Hence the design of these analog and RF components needs a reiteration to comply with the trends of scaling. The hourglass has witnessed momentous research endeavors to mitigate such detrimental effects of scaling on analog and RF performances at various levels of abstraction. The majority of these efforts rely upon the inclusion of calibration circuits to the core at hardware level and/or technology computer-aided design (TCAD) simulation tools at software level. Despite the high degree of precision offered by these approaches, they are not always a handy choice due to the involvement of greater complexity, cost, resource, and design time.

The present analysis attempts to rephrase these predicaments by means of a semi-empirical approach based on device-circuit integration. This amalgamated treatment makes use of mathematical modeling to exploit the device-circuit fundamentals and offers advantages in terms of cost-minimization, promptness, and reliability in obtaining solutions for (almost) any real-time IC design problem with industry acceptable extent of fidelity.

11.1   INTRODUCTORY NOTE

Invention of transistors at the Bell Laboratories in 1947 [1] created possibilities for constructing more advanced circuits based on semiconductor devices that were compact in size and less power-hungry. However, as the design complexity of the circuits increased, newer challenge in the form of size minimization and speed maximization appeared. To meet these challenges, Jack Kilby and Robert Noyce [2] proposed a way for integrating the circuit components on a single silicon wafer. Thus, began the era of integrated circuits or ICs chips for short.

The primitive ICs of the early 1960s could only accommodate as many as 10 semiconductor-based diodes, transistors, resistors, and capacitors on a single Silicon wafer, making it possible to fabricate one or more logic gates on a single 'chip.' This was known as small-scale integration (SSI). Advancements in manufacturing technique in the late-1960s led to ICs with hundreds of logic gates, coined as medium-scale integration (MSI). Further improvements in the 1970s came up with large-scale integration (LSI) circuits having at least a 1,000 logic gates (i.e., tens of thousands of transistors on a single chip). Hundreds of thousands of transistors integrated on a single chip during the 1980s marked the beginning of an era called very-large-scale integration (VLSI).

Following the trends of technology, this number has surpassed millions of gates and billions of transistors per chip in today's IC industry.

11.2   SYSTEM-ON-CHIP

In the year 1965, Gordon E. Moore, prior to founding Intel, proposed the famous Moore's Law for ICs industry. According to him, "the complexity for minimum component costs has increased at a rate of roughly a factor of two per year" [3]. The 1965 version of Moore's Law is graphically represented in Figure 11.1. This proposition was amended to 'every two years' in 1975.

The process of IC manufacturing is typically an amalgamation of several intricate steps that are executed sequentially. In line with the advent of new technologies, each of these steps gets more complex. Consequently, the need of more sophisticated tools/mechanisms become inevitable. Due to the decentralized approach in the design and fabrication of ICs in semiconductor industry, different companies master different segments of the IC manufacturing process. But in order to collaborate their developments, there has to be some sort of a standardized guideline between these companies.

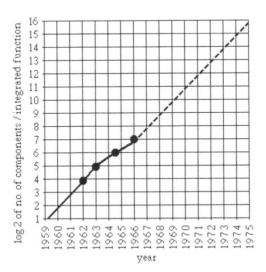

FIGURE 11.1   Moore's law.

International technology roadmap for semiconductors (ITRS) is one such directive from a sanctum of experts, to highlight the trends of technological advancements to suit the requirements of the fabrication industry [4]. ITRS can be regarded as a bona fide aid for these companies to understand the trend in the market evolution and to bring about a parity between demand and supply.

An IC which consolidates various functional integrant of an entire electronic system over a lone substrate is coined as system-on-chip or SoC [5]. This compaction design philosophy makes SoCs more viable than conventional systems when it comes to overall performance or economy. Compared to the conventional microcontrollers, SoCs are more powerful systems as they are capable of driving complete embedded mechanisms. Some of the common application domains of SoCs are cell phones, biomedical devices, consumer electronic appliances, etc. Generally, SoCs are customized in line with their application domain, followed by identification of the relevant integrant based on the desired attributes. Then all the integrant are concatenated within a software development environment using dedicated programming languages to create the entire blueprint. This process is coined as the 'functional design and verification' stage.

System designers have employed various tools to carry out this task of SoC performance assessment and debugging prior to sending them to the foundry. Due to the higher packaging density of modern-day SoCs, this 'design and verification' stage may cause a delay in the production sequence [6].

11.3 STUDY OF DEVICE-CIRCUIT INTEGRATION

The study of device-circuit integration, also known as device-circuit interaction analysis, deals with technology aware circuit and system design methodologies that consider device architecture and technological challenges to achieve inclusive design optimality [7]. To comply with the requirements of present-day consumers, the ever-increasing demand of SoCs has catalyzed the implication of device-circuit integration studies to a greater extent. This discourse takes some of the common challenges into account which are encountered during device-circuit design at different technology nodes and levels of abstraction. A relevant device-circuit topology used in biomedical applications is considered in Section 11.6 to assess and justify the importance of this approach.

11.3.1 DESIGN METRICS

A design space comprises of a pool of various design metrics. In the present context, the phrase 'design space' is used to signify the overall SoC implementation. The term design metric is a measurable feature of a system's implementation [8]. Design metrics typically compete with one another; improving one may lead to worsening of the other. This phenomenon can be collated to a wheel with numerous pins $(M_1, ..., M_n)$, as illustrated in Figure 11.2; where the wheel represents the design space and every pin is analogous to a design metric. Now, pushing a pin further may cause some of the pins to pop out and some to pop in. For example, if the size of an SoC is shrunk, then the power consumption would decrease, but the overall design complexity would increase. Similarly, in order to deliver higher throughput (i.e., to speed-up the SoC), the overall design complexity would increase; leading to greater power consumption. Therefore, it is evident from the example that the SoC design space is packed with stifle design metrics, which needs judicious handling.

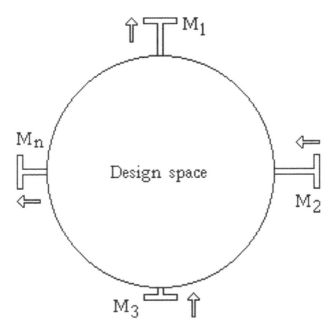

FIGURE 11.2   Competing design metrics.

11.3.2   ANALOG AND RF METRICS

The presence of analog and RF integrant in SoCs facilitate the reciprocation of data with this external world via various modes of communication. The semiconductor industry continues to challenge analog and RF IC designers with its escalating demand for higher performance and better compatibility amidst the digital surge of CMOS processes [9]. Consequently, analog, and RF circuits entail a 'trade-off packed' multidimensional design space. The true severity of these trade-offs can only be intercepted by relevant technological characterization; with judicious modeling as the key to this casket. While efforts toward improving sub-micron device models continues vigorously, aggressive trends of scaling keep affecting the performance of analog and RF metrics.

The hourglass has witnessed momentous research endeavors to miti-gate such detrimental effects of scaling on analog and RF performances at various levels of abstraction by inclusion of calibration circuits to the core at the hardware level and/or technology computer-aided design (TCAD) simulation tools at software level. However, both the approaches have their inherent shortcomings:

- Due to the lack of universally-applicable benchmarks, many test structures must be built to meet the needs of various subsystems;
- Some device characteristics are comparatively difficult to measure. Thus, proper calibration circuit must be included on the die;
- Some of the measured features are difficult to incorporate in simulations;
- Number of test structures involve substantial amount of design time and effort;
- Portability of testing mechanisms to the successive technological generation with minimum modification is also a major concern.

Some of the common device and circuit attributes of interest in analog and RF design are: AC and DC behavior, linearity, matching, noise, and temperature dependence, etc. It is observed, these design metrics can be greatly simplified if the measured data points describing the behavior of device and sub-circuits are obtained [9].

11.4   MOTIVATION

There is a volume of research work on computer aided design (CAD) of analog devices and circuits, which dates back more than 20 years. To sum them up, it can be inferred that the performance measures of an IC is influenced by a series of interrelated design metrics. The job of an IC designer is to harness these design metrics in order to optimize the overall performance of a system. This optimization task is coined by the IC manufacturers as a design problem, which can be approached in a multitude of ways to identify the best implementation for any given application and constraints.

The aim of the present study is to explore the possibility of developing a mathematical model for optimization of such design problems. The design problem discussed in Section 11.6 comprises of a fundamental device and circuit optimization instance related to SoC applications; viz., biomedical CMOS low-noise amplifier. For the aforesaid design problem, the objective is to determine a low-noise and low-power solution, subject to constraints on some of the key device and/or circuit characteristics such as AC and/or DC behavior, linearity, matching, noise, and temperature, etc. In a nutshell, finding the most optimum solution conforming to a user-defined design space remains as the common design goal for this trade-off packed design problem. The motivation of this study is to meet these by the use of mathematical modeling techniques.

Use of mathematical modeling in the field of electronic device-circuit optimization is in practice for quite some time now; which has produced some of well-known techniques; viz., classical optimization, knowledge-based optimization, global optimization, etc. Salient pros and cons of these optimization techniques are discussed later in this section. Depending on the type of design problem and the optimization technique used, some of the usual shortcomings encountered are listed as follows:

- Sloppy execution speed for large-scale problems;
- Dependence of optimum solution on initial design settings;
- Tendency of getting stuck at a local optimal solution;
- Resource hungry and cumbersome computation, etc.

These shortcomings, along with other inherent limitations of the existing processes are easily surpassed by the promptness and efficiency of geometric programming, making it an ideal choice for this study. It has the ability to determine globally optimum solutions to electronic device-circuit optimization problems with least human oversight.

It is noteworthy that integration of mathematical modeling to standard TCAD toolsets is yet to be carried out extensively. As a result, presently, all research efforts pertaining to mathematical modeling of electronic devices and circuits are limited to either instance specific coding or user-defined functions/toolboxes; which can be used alongside standalone mathematical scripting engines such as MATLAB. However, with the growing interest in this field, the chance of using mathematical modeling as a mainstream TCAD optimization tool appears potent.

11.4.1 SCOPE OF WORK

As discussed earlier; the growing demand of integrating analog, digital, and mixed-signal components in modern-day SoCs face serious bottlenecks due to the comparatively immature design automation processes of analog integrant. However, the trends of efficient analog IC design rely on the development of robust circuit simulation tools, to ascertain the desired functionality of ICs prior to fabrication. Aspects of IC design automation tools include system synthesis, symbolic analysis, layout planning, functionality testing and optimization to meet the desired performance specifications. Profound discussions on CAD of analog ICs can be found in existing literature [10–12].

The early CAD tools featured simulation and netlist-based circuit description languages as its mainstay (such as SPICE). Whereas, newer generation tools focused on system-level modeling; based on analog hardware description languages (AHDLs) [13, 14]. But, due to the lack of standardization; AHDLs fail to bring coherence between various steps of analog IC design. For fabrication technology to thrive, these newer generation tools should function like a unifying factor across the entire design hierarchy with provisions to handle both circuit-level and system-level models. While analog IC design tools are oriented towards full-custom designs, similarly quickly-implementable design techniques are crucial for semi-custom structures to determine suitable design specifications. To correlate the design efforts, it is customary to emphasize on some of the key concerns of VLSI circuit design [15].

In the realm of application-driven SoCs; low-power signal processing of scaled-down technologies being one of the prime concerns, deserves sort of special treatment to manufacture high-performance VLSI circuits. But, due to limitations on time and other resources; conventional CAD tools are not always a handy choice for system performance prediction, design functionality test and trade-off analysis of SoCs prior to fabrication. Factors such as ease of operation, prompt throughput and meager resource utilization supersede the necessity to maintain the highest degree of accuracy at this stage.

This is where mathematical modeling comes to aid. Salient attributes of some of the well-known mathematical modeling techniques such as; classical optimization, knowledge-based optimization, global optimization, convex optimization, and geometric programming are briefly delineated in the remainder of this section.

11.4.2   *CLASSICAL OPTIMIZATION*

Classical optimization methods, viz., Lagrange multiplier (finding the local maxima and minima of a function subject to equality constraints), sequential quadratic programming (iterative method for constrained nonlinear optimization) and steepest descent (finding contour integral in the complex plane to pass near a stationary/saddle point) are extensively used in analog device/circuit CAD; which is reported in literature [16]. Some of the most widely used general-purpose optimizers, such as LANCELOT (large and nonlinear constrained extended Lagrangian optimization technique) [17], MINOS (Modular In-core Nonlinear Optimization System) [18] and NPSOL (nonlinear programming systems optimization laboratory) [19] can be found in previous literature [20–23].

Other approaches based on classical optimization and its extensions such as a minimax formulation (a decision rule used in game theory and statistics for minimizing the possible loss in a worst-case/maximum loss scenario), are included in literature [24–31]. Classical optimization methods are compatible with complex circuital models; which consist full SPICE simulations at each iteration level [32]. The main benefit of using these methods lies in the wide range of problems that can be handled; with the sole requirement of performance measures and/or their derivatives. The main disadvantage of the classical optimization methods is that they can estimate locally optimal designs only. This implies that the design is as good as its neighboring designs at the least, i.e., minor alterations in parameters may terminate as an inferior solution or a completely infeasible one. Due to this reason, globally optimal result cannot to ascertained.

The same problem arises while determining feasibility using a classical optimization method; it may not always identify a feasible design, even though one exists. These optimization methods have a knack of getting stuck at local minima. This blemish is so well known that it is often not stated in the chapters. Usually, treating non-global solutions using these methods commences with the minimization of the initial settings (one after the other) and terminates with the selection of the best result available after minimization. However, this technique can only improve the probability of detecting global optimum at most. Due to the fact that computation effort gets multiplied by the number of different initials designs tried, this approach impacts one of the prime merits of classical methods, i.e., execution speed. Even procurement of good initial designs requires a significant amount of time and human oversight.

11.4.3   KNOWLEDGE-BASED OPTIMIZATION

Knowledge-based optimization methods are well known alternatives to classical optimization methods in analog device/circuit CAD. Examples include genetic algorithms (a metaheuristic inspired by the natural selection process, belonging to a larger class of evolutionary algorithms) and evolution systems (a type of system, which reproduce with mutation where the fittest elements survive and the less fit die down) by Ning et al. [33] and others [34, 35]; fuzzy logic-based systems (a mathematical system that analyzes analog inputs in terms of continuous logical variables between 0 and 1, in contrast to discrete-valued digital logic) [36, 37]; and systems based on special heuristics by Degrauwe et al. [38, 39] and others [40–42].

Prime advantage offered by these methods is that they do not use derivatives as the functional basis, consequently there are few limitations on the types of performance measures, problems, and specifications that can be handled using these methods. Some of the demerits, encountered by these approaches are as follows: rather than finding a global optimum, they have the tendency to settle with a local one; that too the outcome and its precision depend largely on the initial settings; even the design and training drills associated with these methods consume substantial human oversight and time.

11.4.4 GLOBAL OPTIMIZATION

Such optimization methodologies which can ascertain a global optimum design are in use for quite some time in analog device/circuit CAD. Branch and bound (an estimation of the lower and upper bounds of a region/branch of the search space and approaches exhaustive enumeration as the size of the region tends to zero) [43] and simulated annealing (SA) (from its inspiration lying with equilibration in metallurgy it is a probabilistic technique for approximating the global optimum of a given function) [4, 45] are some of the popular algorithms based on global optimization techniques. At every level of execution, branch, and bound algorithm applies a lower limit on the design metrics to keep the design-space under check. This approach is needed to ensure a valid outcome; in other words, branch and bound algorithm can produce global optimum solution subject to predefined tolerances. A severe disadvantage of this approach lies in the fact that the computation payload rises steeply with the problem size; turning this method into a very sloth one while dealing with large-scale problems.

SA, unlike other global optimization methodologies, can overcome the tendency of clogging at local minima. Although this technique can theoretically determine the global optimum solution, but there is no certainty as such during execution. It is due to the fact that the annealing modules function correctly under ideal scenarios only (which is bound to differ during implementation). Similar to classical and knowledge-based optimization methods, SA can handle different types of variables and design metrics; but unlike the previous two, there are no real-time learning modules readily available here. Several tools, which make use of the SA algorithm are reported in Refs. [46–50]. Advantage of SA is that it can efficiently handle sparsity and reduces the probability of finding a local solution. Sloppy execution speed and the inability to determine global solution under all circumstances are identified as its demerits.

11.4.5 CONVEX OPTIMIZATION AND GEOMETRIC PROGRAMMING

Mathematical attributes of convex optimization [51] and geometric programming [52] have been practiced and appreciated for decades. Convex optimization techniques concerning the minimization of convex functions over convex sets are simpler than the general optimization methods, since global minimum replaces local minimum. However, practical applications of these methodologies have started seeing broad daylight only after the deployment of interior-point algorithms. They are ideally suited for larger problems. Regardless to the primary settings, these techniques can achieve a global optimum outcome with unprecedented efficiency. Another merit of is that if the design requirements are too tight to be complied simultaneous, i.e., if they are mutually inconsistent, then the program explicitly reports the same by raising a flag. Also, due to the absence of 'learning modules,' there is no unnecessary delay in the execution process.

A major drawback of this method is that the types of compatible functions (used for defining various metrics) are very slender in comparison to the other approaches described earlier; which is eventually a trade-off to accommodate efficiency and promptness in finding truly global solutions. Based on these pros and cons, geometric programming is opted in this study to estimate optimum performance metrics. Apart from the area of research discussed in this chapter; implications of geometric programming can be observed in semiconductor device sizing in digital circuits [53] and other problems reported by Sapatnekar et al. [54, 55] and Shyu et al. [56]. Usage of convex optimization is also observed in several prior literature concerning small-scale device/circuit design problems [20, 57–62].

11.4.6 OBSERVATIONS

Based on the erstwhile survey of literature, it can be postulated that orthogonal convex optimization based geometric programming offers a string of merits over other optimization methods:

- Efficient handling of large-scale design problems;
- Guaranteed global optimal design solution;
- Prompt throughput with least human oversight;
- Provision for design feasibility and trade-off analysis.

Which makes geometric programming the preferred technique for the design instance in Section 11.6.

11.5 INTRODUCTION TO GEOMETRIC PROGRAMMING

Geometric programs (GP) [51, 52, 63] are a special bracket of problems with an objective to be minimized mathematically; subject to predefined constraints. Newly developed GP methodologies can solve large scale practical problems related to semiconductor device and circuit design with efficiency and ease. The primary idea of GP modeling is to express any practical problem in a compatible manner either by exact formulation or by approximation. Although, successful formulation of every design aspect cannot be guaranteed, but any duly formulated problem can be solved efficiently and reliably using this method. Remainder of this section consolidates the fundamentals of GP modeling.

11.5.1 STANDARD REPRESENTATION

GP is the family of optimization problems represented by the generic equation:

Optimize	$f_0(x)$		
Subject to	$f_i(x) \leq 1,$	$i = (1,...,m)$	(11.1)
	$g_i(x) = 1,$	$i = (1,...,p)$	
	$x_i > 0,$	$i = (1,...,n)$	

where; $f_1, ..., f_m$ are posynomial inequality constraints; and $g_1, ..., g_p$ are monomial equality constraints of vector x comprising of n real positive variables x_1, ..., x_n. The objective function f_0 is either a posynomial to minimize (only) or a monomial to maximize/minimize (scenario specific).

11.5.2 MONOMIAL AND POSYNOMIAL FUNCTIONS

A monomial [51, 52] function $g(x)$: $\mathbf{R}^n$ is defined by a vector comprising of only positive variables. It is represented as follows:

$$g(x) = cx_1^{a_1}...x_n^{a_n} \tag{11.2}$$

where; $c > 0$ is the coefficient; and vector $a = a_1$, ..., a_n is the exponent of the monomial. Any variable or positive constant can be regarded as a monomial as well.

When a monomial is divided and/or multiplied by any positive quantity and/or raised by any exponent; its property remains unchanged.

A posynomial [51, 52] function $f(x)$: $\mathbf{R}^n$ is defined as a sum of vectors comprising of positive variables only. It is represented as follows:

$$f(x) = \sum_{k=1}^{K} c_k g_k(x) \tag{11.3}$$

where; $g_k(x)$ are monomials; and $c_k \geq 0$ for $k = 1$, ..., K.

When a posynomial is added and/or divided and/or multiplied by any positive quantity or any other monomial; its property remains unchanged.

11.5.3 EXECUTION PROCESS

An interior-point algorithm is used for executing the GP. Handling a problem of this sort is a two-step process; the first step commences with conversion of all constituting functions into convex form, which is done by taking their 'logarithm'; the second step is to execute these reiterated functions in the form of a single optimization problem.

Each reiterated function is represented either by an equality (==) or an inequality (<=). This two-step GP execution process is capable of finding the globally optimal solution without the need of any parameter tuning or initial guess. The reiterated problem formulation is given as follows:

Optimize	$f_0(y) = \log\left(\sum_{k=1}^{K_0} e^{a_{0k}^T y + b_{0k}}\right)$	
Subject to	$f_i(y) = \log\left(\sum_{k=1}^{K_0} e^{a_{ik}^T y + b_{ik}}\right) \leq 0, \quad i = (1,...,m)$	(11.4)
	$g_i(y) = a_i^T y + b_i = 0, \quad i = (1,...,p)$	

where; the new variables $yi = \log xi$, such that $xi = e^{yi}$.

11.5.4   TEST OF FEASIBILITY, TRADE-OFF, AND SENSITIVITY

As stated in Section 11.5.3, the approach of execution of any GP is a two-step process: a test of mutual consistency of constraints is carried out in the first step (test of feasibility) [3]. If mutual consistency is observed, then an attempt of finding an optimum solution is made in the second step.

Under practical scenario, infeasibility means constraint specifications are too stringent to be complied simultaneously; and one constraint at the least needs to be relaxed. When a problem is identified as infeasible, GP does the necessary flagging.

During trade-off testing; GP constraints are varied to observe their effects on the optimal value [3]. This reflects the fact that the constraints in a practical problem may vary for compelling reasons. From the standard representation of GP in Eqn. (11.1), an unsettled GP is formulated for trade-off test:

Optimize	$f_0(x)$		
Subject to	$j = (1,...,m)$	$j = (1,...,m)$	(11.5)
	$g_j(x) = r_j,$	$j = (1,...,p)$	

where; q_j and r_j are positive valued quantities. Let the function $t(q, r)$ represent the optimum result for the unsettled formulation in Eqn. (11.5).

This function is examined for different q and r during trade-off test. For $q_j > 1$, the jth inequality function is loosened by $100(q_j - 1)\%$; and for $q_j < 1$, the jth inequality function is tightened by $100(1 - q_j)\%$. Similarly, the context of jth equality constraint can be interpreted with respect to r_j.

Test of sensitivity [3] shares a proximal relation with trade-off test. Influence of differential changes in constraints upon optimal objective is scrutinized here, i.e., $t(q, r)$ for q_j and r_j in the vicinity of '1' is considered. Surmising that the optimum value of objective $t(q, r)$ can be differentiated at $q_j = 1$, $r_j = 1$; changes in its values with respect to very small changes in q_j and r_j can be foresighted using the plot obtained from trade-off test. This notion is given by the following derivative functions:

$$\left.\frac{\partial t}{\partial q_j}\right|_{q=1,r=1} \quad \text{and} \quad \left.\frac{\partial t}{\partial r_j}\right|_{q=1,r=1} \tag{11.6}$$

Small variation in optimum solution caused by a small variation in j^{th} inequality can be seen as the sensitivity of jth inequality, it is given by:

$$O_j = \left.\frac{\partial \log t}{\partial \log q_j}\right|_{q=1,r=1} = \left.\frac{\partial \log t}{\partial q_j}\right|_{q=1,r=1} \tag{11.7}$$

Small variation in optimum solution caused by a small variation in j^{th} equality can be seen as the sensitivity of jth equality, it is given by:

$$L_j = \left.\frac{\partial \log t}{\partial \log r_j}\right|_{q=1,r=1} = \left.\frac{\partial \log t}{\partial r_j}\right|_{q=1,r=1} \tag{11.8}$$

It is customary to use these normalized derivatives (O_j and L_j), which describes the variation in optimum solution for a variation in q_j and r_j.

11.5.5 GENERALIZED GEOMETRIC PROGRAM

GGP [52] is a design problem given by Eqn. (11.1). The only difference is that the objective and/or constraints are monomials and/or generalized posynomials. GGPs can be automatically transformed to equivalent GPs by a parser and can be solved with reliability and efficiency.

Positive fractional power, pointwise maximum of posynomials is regarded as generalized posynomials. There is also a possibility that a posynomial equality constraint may exist in a GGP. Such an occurrence can be handled straight away by replacing the posynomial equality with an inequality. This approach is called as constraint relaxation.

11.5.6 SOFTWARE PACKAGE

The *ggplab* version 1.00 is a software package for specifying and solving GPs and GGPs [51, 52]. It is a MATLAB based toolbox distributed under GNU General Public License. *ggplab* comprises of:

- A library of relevant objects used for specifying the optimization problem;
- A two-step problem solving mechanism *gpcvx* based on interior-point algorithm;
- A parser *gpposy* to handle posynomials.

Some caveats associated with *ggplab*:

- Object manipulation overhead of larger problems may affect the problem execution speed;
- *gpcvx* cannot handle sparsity in very large-scale problems;
- *ggplab* cannot handle duality. However, *gpcvx* does.

As an alternative to *ggplab* version 1.00 there is another MATLAB based generic toolbox named CVX, which is capable of handling large and complex GP optimization problems. To impart simplicity in the present analysis, *ggplab* remains as the preferred toolbox.

11.5.7   *OPTIMIZATION PROCESS FLOW*

Salient benefits of using GP over other optimization techniques (for electronic device and circuit optimization) are summarized in Sections 11.4.5 and 11.4.6. Based on those advantages, let us interpret the GP optimization process. Depending on how circuital performances and constraints are evaluated, any optimization technique can be broadly classified into two categories [51]:

- Simulation-basedoptimization techniques; and
- Equation-based optimizationtechniques.

Being an equation-based optimization technique and from the parlance of electronic device and circuit optimization; GP is summarized as a flowchart in Figure 11.3—a set of device and circuit equations, describing an electronic system are considered at the onset, then a group of operational scenarios are imposed on these equations as constraints alongside some of the scenario independent design variables. Then concurrent evaluation of these equations is carried out using *ggplab* toolbox to determine a globally optimum solution (if any), otherwise infeasibility is reported.

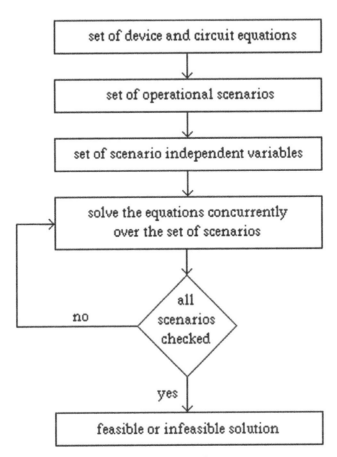

FIGURE 11.3   Flowchart of device circuit optimization process.

The stated approach is exercised in this literature accompanied by a pristine modeling strategy to optimize various performance metrics of an SoC sub-circuit suitable for biomedical application.

Coherence between GP and device-circuit integration comes from the fact that a vast majority of characteristic equations used to define electronic systems can either be directly formulated as convex functions or they can be easily transformed into an analogously by the dint of approximations; making them suitable for GP. Consequently, the efficiency and promptness of GP can be brought into action to determine globally optimal solutions unambiguously, which otherwise is a very cumbersome task while using conventional TCAD tools.

11.6   OPTIMIZATION OF BIOMEDICAL CMOS LOW-NOISE AMPLIFIER

There has been an upsurge in the design of neural recording interfaces for biomedical measurements [64–66]. These interfaces play a significant role in clinical and neuroscience applications for capturing vital health information related to the human body for efficient diagnosis [67, 68]. These systems comprise of microelectrode(s) for biomedical activity capturing, followed by low-noise amplifiers (LNAs) for signal conditioning and a mixed-signal circuitry for digitization and processing of the acquired data. LNA is a key element in these structures, which must be capable of boosting the feeble signals detected by the microelectrodes and percolate out the unnecessary parts. Various proposals are given for solving this challenging noise-power-area trade-off packed biomedical LNAs [69–75].

The optimization strategy based on GP and all-inversion region MOSFET, is used here to implement the CMOS LNA, with discernible changes to prior endeavors [76, 77].

11.6.1   LOW-NOISE AMPLIFIER TOPOLOGY

The CMOS LNA topology under consideration is of capacitive feedback network (CFN) [67, 69, 74] type, as shown in Figure 11.4.

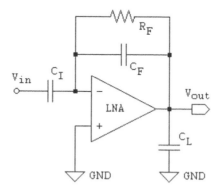

FIGURE 11.4   CFN LNA topology.

This amplifier topology can be approximated by a three-pole network transfer function $H(s)$ [78, 79].

$$H(s) \approx \frac{A_v}{\left(1 + \dfrac{s}{p_L}\right)\left(1 + \dfrac{s}{p_H}\right)\left(1 + \dfrac{s}{p_C}\right)}$$

(11.9)

where; p_L is dominant pole; p_H is output pole; p_C is compensation pole; and A_v is the open-loop DC voltage gain.

The open-loop preamplifier circuit used for implementing this neural data acquisition interface is shown in Figure 11.5.

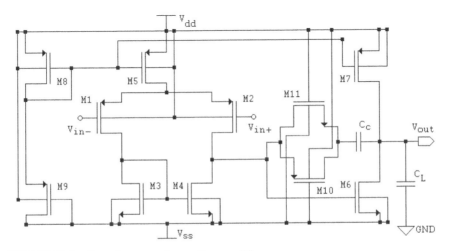

FIGURE 11.5   Open-loop neural recording amplifier.

This amplifier must exhibit low power, low input-referred noise, and high common-mode rejection ratio to conform with the target biomedical signal monitoring domain. Such signals are characterized by their low-frequency and low-voltage levels [80–83] given in Table 11.1.

TABLE 11.1   Amplitude and Frequency of Biomedical Signals

Parameter	ECG	EEG	EMG	Unit
Frequency	0.05–250	0.5 – 200	$0.01–10\times10^3$	Hz
Amplitude	$5\times10^{-6}–8\times10^{-3}$	$2\times10^{-6}–200\times10^{-6}$	$50\times10^{-6}–10\times10^{-3}$	V

The amplifier in Figure 11.5 has two stages with a Miller capacitor or compensation capacitor (C_c) and a nulling resistor (R_z). The first stage is a PMOS differential input pair and the second, a gain stage. Low frequency and amplitude of the target biomedical signals gives rise to flicker noise in the structure. Spectral density of flicker noise is inversely proportional to the length-width product of MOSFETs, i.e., it has a lesser effect on bigger devices. However, bigger dimensions lead to higher power dissipation. Such countervailing aspects need to be handled judiciously.

Transistors M1, M2, M5, M7, M8, M9, and M10 are PMOS and the rest are NMOS in Figure 11.5. Current source (I_{ref}) is built by shorting the gate and drain of M9. This connection ensures saturation region operation for M9. The nulling resistor (R_z) is formed by shunting M10 with M11. The gates of M10 and M11 are biased at V_{dd} and V_{ss}, respectively. They are illustrated in Figure 11.6.

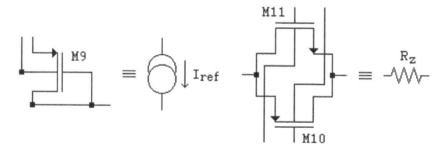

FIGURE 11.6   Active resistors.

Such realization of R_z in the form of an active component; based on a MOSFET pair can significantly improve the dynamic range of the LNA over a single-transistor based variant [78]. The equivalent resistance of R_z can be estimated from Eqn. (11.10).

$$
\begin{aligned}
R_{M10} &= \frac{1}{\mu_{M10} C_{ox} \left(V_{gs} - \left| V_{th,M10} \right| \right)} \\[2mm]
R_{M11} &= \frac{1}{\mu_{M11} C_{ox} \left(V_{gs} - \left| V_{th,M11} \right| \right)} \\[2mm]
R_z &= R_{M10} \| R_{M11}
\end{aligned}
\tag{11.10}
$$

where; R_{M10} and R_{M11} are equivalent resistances for transistors M10 and M11. Also, μ, C_{ox}, V_{gs}, V_{th} along with the respective MOSFET subscript; carry their usual meanings.

11.6.2   FORMULATION OF NEURAL RECORDING AMPLIFIER

This neural recording amplifier is optimized using GP with an objective of minimizing the power dissipation (P); subject to constraints on open-loop DC voltage gain (A_v), common-mode rejection ratio (C_{mrr}), gain-bandwidth (G_{bw}), slew rate (S_r), phase margin (P_m), input-referred noise power spectral density $S_{in}(f)$.

A_v can be expressed in a GP compatible form as:

$$A_v = \frac{2\left(\dfrac{g_m}{I_d}\right)_{M2}\left(\dfrac{g_m}{I_d}\right)_{M6} I_{d,M2} I_{d,M6}}{\left(\left(\dfrac{I_d}{V_A}\right)_{M2} + \left(\dfrac{I_d}{V_A}\right)_{M4}\right)\left(\left(\dfrac{I_d}{V_A}\right)_{M6} + \left(\dfrac{I_d}{V_A}\right)_{M7}\right)} \tag{11.11}$$

where V_A is the early voltage. This inverse posynomial expression of A_v in Eqn. (11.11) is constrained by a minimum gain requirement (A_{v-min}).

Common mode rejection ration is modeled as:

$$C_{mrr} = \frac{2\left(\dfrac{g_m}{I_d}\right)_{M1}\left(\dfrac{g_m}{I_d}\right)_{M3} I_{d,M1} I_{d,M3}}{\left(\left(\dfrac{I_d}{V_A}\right)_{M1} + \left(\dfrac{I_d}{V_A}\right)_{M3}\right)\left(\dfrac{I_d}{V_A}\right)_{M5}} \tag{11.12}$$

C_{mrr} in Eqn. (11.12) is inverse posynomial; constrained by minimum CMRR requirement ($C_{mrr-min}$).

Gain-bandwidth is well approximated by the following expression:

$$G_{bw} \approx \frac{1}{C_c}\left(\frac{g_m}{I_d}\right)_{M1} I_{d,M1} \tag{11.13}$$

G_{bw} in Eqn. (11.13) is monomial; which is constrained by minimum gain-bandwidth (G_{bw-min}).

Considering a simplified realization of R_z based on single MOSFET:

$$R_z = 1\left/\left(\frac{g_m}{I_d}\right)_{M6} I_{d,M6}\right. \tag{11.14}$$

Dominant pole p_L in Eqn. (11.9) is expressed as follows:

$$P_L = \frac{1}{A_v C_c} \left(\frac{g_m}{I_d}\right)_{M1} I_{d,M1} \tag{11.15}$$

Output pole p_H in Eqn. (11.9) is expressed as follows:

$$P_H = \frac{C_c \left(\frac{g_m}{I_d}\right)_{M6} I_{d,M6}}{C_{s1}C_c + C_{s1}C_{s2} + C_c C_{s2}} \tag{11.16}$$

where; C_{s1} is the capacitance at the output node of the first stage and C_{s2} is the capacitance at the output node of the second stage. They are expressed as follows:

$$C_{s1} = C_{gs,M6} + C_{db,M2} + C_{db,M4} + C_{gd,M2} + C_{gd,M4}$$
$$C_{s2} = C_L + C_{db,M6} + C_{db,M7} + C_{gd,M6} + C_{gd,M7} \tag{11.17}$$

where; C_{gs} is the gate to source capacitance; C_{db} is the drain to body capacitance; C_{gd} is the gate to drain capacitance of respective MOSFETs; and C_L is the load capacitance.

Compensation pole p_C in Eqn. (11.9) is expressed as follows:

$$P_C = \frac{1}{C_{s1}} \left(\frac{g_m}{I_d}\right)_{M6} I_{d,M6} \tag{11.18}$$

Slew rate S_r is expressed as an inverse posynomial expression:

$$S_r = \min\left\{\frac{2I_{d,M1}}{C_c}, \frac{I_{d,M7}}{C_c + C_{s2}}\right\} \tag{11.19}$$

The expression in the right-hand side of Eqn. (11.19) has two parts. Therefore, each part has to be treated distinctly with minimum S_r constraint (S_{r-min}).

At unity-gain frequency, phase shift contribution of dominant pole does not exceed 90°. Minimum phase margin constraint (P_{m-min}) is well approximated by the following polynomial inequality:

$$\tan^{-1}\left(\frac{G_{bw}}{p_H}\right) + \tan^{-1}\left(\frac{G_{bw}}{p_C}\right) \le 90° - P_{m-min} \tag{11.20}$$

This phase margin constraint in Eqn. (11.20) cannot be handled by GP straightaway. Using the concept of parameter fitting; $\tan^{-1}(x) \approx 0.75 x^{0.7}$ gives a fit of $\pm 3°$ for angles between $0°$ and $60°$.

Spectral density of input-referred flicker noise at frequency f ($<$ 3-dB frequency) is well approximated by the following expression:

$$S_{in}(f)^2 = 2\left[S_{M1}(f)^2 + \left\{ S_{M3}(f) \left(\frac{g_m}{I_d}\right)_{M1} \left(\frac{g_m}{I_d}\right)_{M3} I_{d,M1} I_{d,M3} \right\}^2 \right]$$ (11.21)

where;

$$S_{M1}(f) = \frac{K_{f,M1}}{C_{ox}^2 \left(WLf^{A_f}\right)_{M1}} \qquad S_{M3}(f) = \frac{K_{f,M3}}{C_{ox}^2 \left(WLf^{A_f}\right)_{M3}}$$ (11.22)

where; parameter K_f represents the operating value of flicker noise factor; and A_f is the slope of flicker noise power spectral density for respective MOSFETs. The posynomial in Eqn. (11.21) is constrained by maximum input-referred flicker noise power spectral density $S_{in}(f)_{max}$.

Power dissipation, being the objective function to be minimized, is expressed as:

$$P = V_{dd}\left(I_{d,M5} + I_{d,M7} + I_{d,M8}\right)$$ (11.23)

Biasing voltages, gate overdrive, signal swing, symmetry, and matching, and systematic input offset constraints are also included in the problem definition.

The optimization problem is summarized in Eqn. (11.24).

Minimize	P	
Subject to	$L_{min} \le L_i \le L_{max}, \; W_{min} \le W_i \le W_{max}, \qquad i = (M1,\ldots,M8)$	
	$\left(\frac{S_{f_i} U_T}{2}\right)^2 \left(\frac{g_m}{I_d}\right)_i^2 C_{inv_i} + S_{f_i} U_T \left(\frac{g_m}{I_d}\right)_i \le 1,$	
	$V_{od} \le \sqrt{2 I_{d,i} \Big/ \mu_i C_{ox} \left(\dfrac{W}{L}\right)_i},$	
	$A_{v-min} \le A_v, \; C_{mrr-min} \le C_{mrr}, \; G_{bw-min} \le G_{bw}, \; S_{r-min} \le S_r,$	

$$
\begin{aligned}
&S_{in}(f)^2 \leq S_{in}(f)^2_{max}, I_{d,M5} = \frac{(W/L)_{M5}}{(W/L)_{M8}} I_{d,M8}, I_{d,M7} = \frac{(W/L)_{M7}}{(W/L)_{M8}} I_{d,M8}, \\
&0.75\left[\left(\frac{G_{bw}}{P_H}\right)^{0.7} + \left(\frac{G_{bw}}{P_C}\right)^{0.7}\right] + P_{m \cdot min} \leq 90, \frac{(W/L)_{M3}}{(W/L)_{M6}} = \frac{(W/L)_{M4}}{(W/L)_{M6}} = \frac{1}{2}\frac{(W/L)_{M5}}{(W/L)_{M7}}, \\
&L_{M1} = L_{M2}, L_{M3} = L_{M4}, L_{M5} = L_{M7} = L_{M8}, W_{M1} = W_{M2}, W_{M3} = W_{M4}, \\
&\left(\frac{g_m}{I_d}\right)_{M3} = \left(\frac{g_m}{I_d}\right)_{M4} = \left(\frac{g_m}{I_d}\right)_{M6}, \left(\frac{g_m}{I_d}\right)_{M5} = \left(\frac{g_m}{I_d}\right)_{M7} = \left(\frac{g_m}{I_d}\right)_{M8}, \\
&\left(\frac{g_m}{I_d}\right)_{M1} = \left(\frac{g_m}{I_d}\right)_{M2} = \frac{1}{2}\left(\frac{g_m}{I_d}\right)_{M5}, \\
&I_{d,M1} = I_{d,M2}, I_{d,M3} = I_{d,M4}, \ I_{d,M1} = I_{d,M3} = 0.5 I_{d,M5}, I_{d,M6} = I_{d,M7}, \ 0.22 C_L \leq C_c a
\end{aligned}
\tag{11.24}
$$

11.6.3   RESULT INTERPRETATION

Device parameters are taken from a BSIM 3v1 (LTspice level-8) CMOS model for 180 nm technology node at 300 Kelvin; y MOSIS. Constraint specifications are listed in Table 11.2. Minimum S_r constraint would ensure adequate open-loop voltage gain in the first stage and a constant current input to the second stage for linear output. Minimum phase margin constraint would fortify possibilities of any in-phase lag at the LNA output.

TABLE 11.2   Constraint Specifications for Biomedical LNA

Constraint	Specification	Unit
Device width	$1.8 \leq W \leq 180$	μm
Device length	$0.18 \leq L \leq 1.8$	μm
Load capacitance	1	pF
Supply voltage	1.25	V
Oxide thickness	4.1	nm
Open-loop DC voltage gain	≥ 60	dB
CMRR	≥ 60	dB
Gain-bandwidth	≥ 1	MHz
SR	≥ 1	V/μs
PM	≥ 60	degree
Input-referred noise	≤ 450	nV/√Hz
Power	*minimize*	μW

Estimated optimum solution is summarized in Table 11.3.

TABLE 11.3   Optimum Solution for Biomedical LNA

Parameter	Symbol	Optimum Solution	Unit
Device length	$L_{M1} = L_{M2}, L_{M3} = L_{M4}$	1.8	µm
	$L_{M5} = L_{M7} = L_{M8}$	1.43658	
	L_{M6}	1.8	
Device width	$W_{M1} = W_{M2}$	180	µm
	$W_{M3} = W_{M4}$	180	
	W_{M5}	180	
	W_{M6}	20.87414	
	W_{M7}	10.43707	
	W_{M8}	1.43658	
Nulling resistance	R_z	420.7278	kΩ
Compensation capacitance	C_c	1.780275	µA
Open-loop DC voltage gain	A_v	80.0	dB
CMRR	C_{mrr}	82.70868	dB
Gain-bandwidth	G_{bw}	22.44995	MHz
SR	S_r	1	V/µs
PM	P_m	78.07401	degree
Input-referred noise	$S_{in}(f)$	319.6656	nV/√Hz
Power	P	64.8477	µW

In Table 11.3, the nulling resistance R_z between the input and output stages is a crucial parameter, since it suppresses the effect of the Right-Hand Plane Zero in the network transfer function to offer greater stability by maintaining balance between the two stages of the amplifier.

To assess the correctness of the optimum data, key performance measures obtained from this method are compared with that of SPICE simulation data in Table 11.4.

TABLE 11.4   Comparison with SPICE Simulation Data

Parameter	This Method	SPICE	Unit
Open-loop DC voltage gain	80.0	67.363	dB
CMRR	82.70868	94.124	dB
Gain-bandwidth	22.44995	32.959	MHz
SR	1	1.45	V/µs
PM	78.07401	63	degree
Power	64.8477	64.8476	µW

Marginal deviation between the results, observed in Table 11.4, are caused by the approximations made in the design for modeling compatibility.

Optimal device and circuit parameters from Table 11.3 are used for rigging the amplifier circuit in SPICE for AC and DC analysis, to obtain Bode plot and power dissipation curves, as shown in Figures 11.7 and 11.8, respectively.

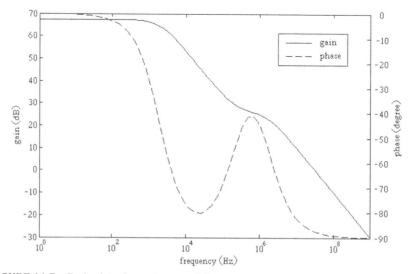

FIGURE 11.7   Bode plot of neural preamplifier.

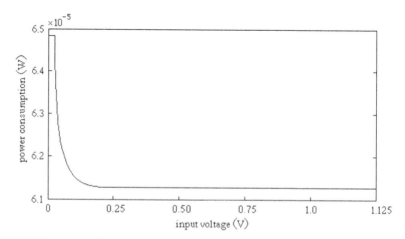

FIGURE 11.8   Power dissipation curve of neural preamplifier.

11.6.4 INFERENCES

Since this amalgamated approach is valid for all three regions of MOSFET inversion; thus, it offers greater flexibility, and it is certainly a reliable approach for predicting optimum behavior of biomedical low noise amplifier.

It is customary to state that apart from the optimum solution enlisted in Table 11.3, additional information on MOSFET operating region and transconductance efficiency can be procured from respective inversion coefficients. According to that data, all transistors (except the active resistors) are operating in moderate inversion; indicating lower power dissipation for this configuration.

Due to limitations on the types of functions that can be handled by GP; parameters such as negative slew-rate, positive power-supply rejection ratio could not be accommodated in this formulation. But the plausibility of incorporating further accuracy by the dint of judicious modeling strategies; signifies the abundant scope of this technique.

KEYWORDS

- **amplifier**
- **biomedical LNA**
- **mathematical modeling**
- **neural preamplifier**
- **neural recording amplifier**
- **system on chip**

REFERENCES

1. Gorton, W., (1998). The genesis of the transistor. *Proc. IEEE 86*, 50–52.
2. Horowitz, P., & Winfield, H., (1989). *The Art of Electronics* (2nd edn.). Cambridge University Press.
3. Gordon, E. M., (1965). Cramming more components onto integrated circuits. *Electronics, 38*(8).
4. Gargini, P., (2000). The international technology roadmap for semiconductors (ITRS): Past, present and future. *22nd GaAs IC Symposium*, 3–5, Seattle, WA, USA, Nov.
5. Bennett, P., & Times, E. E., (2004). *The Why, Where and What of Low-Power SoC Design.*
6. Goering, R., & Times, E. E., (2004). *Is Verification Really 70 Percent?*

7. Roy, K., Kulkarni, J. P., & Gupta, S. K., (2009). Device/circuit interactions at 22nm technology node. *Proc. 46th Annual Design Automation Conference DAC '09*, 97–102. San Francisco, CA, USA.

8. Vahid, F., & Givargis, T., (2009). *Embedded System Design a Unified Hardware/ Software Introduction* (3rd edn.). Wiley-India.

9. Razavi, B., (1998). CMOS technology characterization for analog and RF design. *Proc. IEEE Custom Integrated Circ. Conference*, 3–5. Santa Clara, CA, USA.

10. Antao, B. A. A. (1996). Trends in CAD of analog ICs. *IEEE Circuits Devices Mag., 12*(5), 31–41.

11. Carley, L. R., Gielen, G. G. E. Rutenbar, R. A., & Sansen, W. M. C., (1996). Synthesis tools for mixed- signal ICs: Progress on front-end and back-end strategies. In: *Proc. 33rd Annual Design Automation Conf.* (pp. 298–303). Las Vegas, NV, USA, Jun.

12. Carley, L. R., & Rutenbar, R. A., (1988). How to automate analog IC designs. *IEEE Spectrum, 25*, 26–30.

13. Antao, B. A. A. (1996). AHD Languages: A must for time-critical designs. *IEEE Circuits Devices Mag., 12*(4), 12–17.

14. Gielen, G., Wambacq, P., & Sansen, W. M., (1994). Symbolic analysis methods and applications for analog circuits: A tutorial. *Proc. IEEE, 82*(2), 287–304.

15. Ismail, M., & Fiez, T., (1994). *Analog VlSI: Signal and Information Processing*. International Edition. McGraw-Hill series in electrical and computer engineering.

16. Brayton, R. K., Hachtel, G. D., & Sangiovanni-Vincentelli, A., (1981). A survey of optimization techniques for integrated-circuit design. *Proc. IEEE, 69*, 1334–1362.

17. Conn, A. R., Gould, N. I. M. & PToint, H. L., (1992). *LANCELOT: A Fortran Package for Large-Scale Nonlinear Optimization (Release A)* (Vol. 17). New York: Springer-Verlag.

18. Murtagh, A., & Saunders, M. A., (1983). *Minos 5.4 User's Guide*. Systems Optimization Lab., Stanford Univ., Stanford, CA, Tech. Rept. SOL 83-20R.

19. Gill, P. E., Murray, W., Saunders, M. A., & Wright, M. H., (1986). *User's Guide for NPSOL (Version 4.0): A Fortran Package for Nonlinear Programming*. Operations Res. Dept., Stanford Univ., Stanford, CA, Tech. Rept. SOL 86–2, Jan.

20. Chang, H., et al., (1992). A top-down constraint-driven design methodology for analog integrated circuits. In: *Proc. IEEE Custom Integrated Circuit Conf.* (pp. 8.4.1–8.4.6).

21. Maulik, P. C., & Carley, L. R., (1991). High-performance analog module generation using nonlinear optimization. In: *Proc. 4th Annu. IEEE Int. ASIC Conf. Exhibit.* (pp. T13-5.1–T13-5.2).

22. Maulik, P. C., Flynn, M. J., Allstot, D. J., & Carley, L. R., (1992). Rapid redesign of analog standard cells using constrained optimization techniques. In: *Proc. IEEE Custom Integrated Circuit Conf.* (pp. 8.1.1–8.1.3).

23. Conn, A. R., et al., (1996). Optimization of custom CMOS circuits by transistor sizing. In: *Proc. IEEE Int. Conf. Computer-Aided Design* (pp. 174–180).

24. Heikkilã, P., Valtonen, M., & Mannersalo, K., (1988). CMOS op-amp dimensioning using multiphase optimization. In: *Proc. IEEE Int. Symp. Circuits Systems* (pp. 167–170).

25. Koh, H. Y., Séquin, C. H., & Gray, P. R., (1990). OPASYN: A compiler for CMOS operational amplifiers. *IEEE Trans. Computer-Aided Design* (Vol. 9, pp. 113–125).

26. Jusuf, G., Gray, P. R., & Sangiovanni-Vincentelli, A., (1990). CADICS: Cyclic analog-to-digital converter synthesis. In: *Proc. IEEE Int. Conf. Computer-Aided Design* (pp. 286–289).

27. Leyn, F., Daems, W., Gielen, G., & Sansen, W., (1997). Analog circuit sizing with constraint programming modeling and minimax optimization. In: *Proc. IEEE Int. Symp. Circuits Systems* (vol. 3, pp. 1500–1503).

28. Chadha, R., et al., (1987). WATOPT: An optimizer for circuit applications. *IEEE Trans. Computer- Aided Design* (Vol. CAD-6, pp. 472–479).

29. Madsen, K., Niedseln, O., Schjaer-Jakobsen, H., & Tharne, H., (1975). Efficient minimax design of networks without using derivatives. *IEEE Trans. Microwave Theory Techn., MTT-23*, pp. 803–809.

30. Onodera, H., Kanbara, H., & Tamaru, K., (1990). Operational amplifier compilation with performance optimization. *IEEE J. Solid-State Circuits, 25*, 466–473.

31. Harvey, J. P., Elmasry, M. I., & Leung, B., (1992). STAIC: An interactive framework for synthesizing CMOS and BiCMOS analog circuits. *IEEE Trans. Computer-Aided Design, 11*, 1402–1417.

32. Nye, W., Riley, D. C., Sangiovanni-Vincentelli, A., & Tits, A. L., (1988). DELIGHT. SPICE: An optimization-based system for the design of integrated circuits. *IEEE Trans. Computer- Aided Design, 7*, 501–518.

33. Ning, Z., Mouthaan, T., & Wallinga, H., (1991). SEAS: A simulated evolution approach for analog circuit synthesis. In: *Proc. IEEE Custom Integrated Circuit Conf.* (pp. 5.2.1–5.2.4).

34. Kruiskamp, W., & Leenaerts, D., (1995). DARWIN: CMOS op-amp synthesis by means of a genetic algorithm. In: *Proc. 32nd Annu. Design Automation Conf.* (pp. 433–438).

35. Wójcikowski, M., Glinianowicz, J., & Bialko, M., (1996). System for optimization of electronic circuits using genetic algorithm. In: *Proc. IEEE Int. Conf. Electronics, Circuits Syst.* (pp. 247–250).

36. Hashizume, M., Kawai, H. Y., Nii, K., & Tamesada, T., (1989). Design automation system for analog circuits based on fuzzy logic. In: *Proc. IEEE Custom Integrated Circuit Conf.* (pp. 4.6.1–4.6.4).

37. Torralba, A., Chávez, J., & Ranquelo, L. G. F., (1996). FASY: A fuzzy-logic based tool for analog synthesis. *IEEE Trans. Computer-Aided Design, 15*, 705–715.

38. Deg, R. M. G. R., et al., (1987). IDAC: An interactive design tool for analog CMOS circuits. *IEEE J. Solid-State Circuits, SC-22*, 1106–1115.

39. Deg, R. M. G. R., et al., (1989). Toward an analog system design environment. *IEEE J. Solid-State Circuits, 24*, 1587–1597.

40. Gupta, S. K., & Hasan, M. M., (1996). KANSYS: A CAD tool for analog circuit synthesis. In: *Proc. 9th Int. Conf. VLSI Design* (pp. 333, 334).

41. El-Turky, F., & Perry, E. E., (1989). BLADES: An artificial intelligence approach to analog circuit design. *IEEE Trans. Computer-Aided Design, 8*, 680–692.

42. Harjani, R., Rutenbar, R. A., & Carley, L. R., (1989). OASYS: A framework for analog circuit synthesis. *IEEE Trans. Computer-Aided Design, 8*, 1247–1265.

43. Xinghao, C., & Bushnell, M. L., (1996). *Efficient Branch and Bound Search With Application to Computer-Aided Design*. Norwell, MA: Kluwer.

44. Wong, D. F., Leong, H. W., & Liu, C. L., (1988). *Simulated Annealing for VLSI Design*. Norwell, MA: Kluwer.

45. Van, L. P. J. M., & Aarts, E. H. L., (1987). *Simulated Annealing: Theory and Applications*. Amsterdam, The Netherlands: Reidel.

46. Ochotta, E. S., Rutenbar, R. A., & Carley, L. R., (1996). Synthesis of high-performance analog circuits in ASTRX/OBLX. *IEEE Trans. ComputerAided Design, 15*, 273–293.

47. Medeiro, F., Fernández, F. V., Domínguez-Castro, R., & Rodríguez-Vázquez, A., (1994). A statistical optimization-based approach for automated sizing of analog cells. In: *Proc. 31ˢᵗ Annu. Design Automation Conf.* (pp. 594–597).

48. Gielen, G. G. E., Walscharts, H. C. C., & Sansen, W. M. C., (1990). Analog circuit design optimization based on symbolic simulation and simulated annealing. *IEEE J. Solid-State Circuits, 25,* 707–713.

49. Chávez, J., Aguirre, M. A., & Torralba, A., (1993). Analog design optimization: A case study. In: *Proc. IEEE Int. Symp. Circuits Syst.* (Vol. 3, pp. 2083–2085).

50. Yang, H. Z., Fan, C. Z., Wang, H., & Liu, R. S., (1996). Simulated annealing algorithm with multi- molecule: An approach to analog synthesis. In: *Proc. 1996 European Design & Test Conf.* (pp. 571–575).

51. Boyd, S., & Vandenberghe, L., (2009). *Introduction to Convex Optimization with Engineering Applications* (7ᵗʰ edn.).

52. Boyd, S., Kim, S. J., Vandenberghe, L., & Hassibi, A., (2007). A tutorial on geometric programming. *Springer Optimization and Engineering [Online], 8*(1), 67–127. Available: http://www.stanford.edu/~boyd/papers/pdf/gp_tutorial.pdf (accessed on 21 September 2021).

53. Fishburn, J. P., & Dunlop, A. E., (2003). TILOS: A posynomial programming approach to transistor sizing. In: *Proc. ICCAD '85* (pp. 326–328).

54. Sapatnekar, S., Rao, V. B., Vaidya, P., & Kang, S. M., (1993). An exact solution to the transistor sizing problem for CMOS circuits using convex optimization. *IEEE Trans. Computer-Aided Design, 12,* 1621–1634.

55. Sapatnekar, S. S., (1996). Wire sizing as a convex optimization problem: Exploring the area-delay tradeoff. *IEEE Trans. Computer-Aided Design, 15,* 1001–1011.

56. Shyu, J. M., Sangiovanni-Vincentelli, A., Fishburn, J. P., & Dunlop, A. E., (1988). Optimization-based transistor sizing. *IEEE J. Solid-State Circuits, 23,* 400–409.

57. Maulik, P. C., Carley, L. R., & Rutenbar, R. A., (1995). Integer programming based topology selection of cell-level analog circuits. *IEEE Trans. Computer-Aided Design, 14,* 401–412.

58. Maulik, P. C., Carley, L. R., & Allstot, D. J., (1993). Sizing of cell-level analog circuits using constrained optimization techniques. *IEEE J. Solid-State Circuits, 28,* 233–241.

59. Vassiliou, I., et al., (1996). A video-driver system designed using a top-down, constraint-driven, methodology. In: *Proc. 33ʳᵈ Annu. Design Automation Conf.*

60. Su, H., Michael, C., & Ismail, M., (1994). Statistical constrained optimization of analog MOS circuits using empirical performance tools. In: *Proc. IEEE Int. Symp. Circuits Systems* (Vol. 1, pp. 133–136).

61. Vandenberghe, L., Boyd, S., & El Gamal, A., (1998). Optimizing dominant time constant in RC circuits. *IEEE Trans. Computer-Aided Design, 2,* 110–125.

62. Vandenberghe, L., Boyd, S., & El Gamal, A., (1997). Optimal wire and transistor sizing for circuits with nontree topology. In: *Proc. 1997 IEEE/ACM Int. Conf. Computer-Aided Design* (pp. 252–259).

63. Duffin, R., Peterson, E., & Zener, C., (1967). *Geometric Programming—Theory and Application.* Wiley, New York.

64. Sodagar, A. M., Perlin, G. E., Yao, Y., & Najafi, K., (2009). An implantable 64-channel wireless microsystem for single-unit neural recording. *IEEE J. Solid-State Circ., 44,* 2591–2604.

65. Harrison, R. R., et al., (2009). Wireless neural recording with single low-power integrated circuit. *IEEE Trans. Neural Sys. & Rehab. Eng., 17*, 322–329.

66. Chae, M. S., Yang, Z., Yuce, M. R., Hoang, L., & Liu, W., (2009). A 128-channel 6 MW wireless neural recording IC with spike feature extraction and UWB transmitter. *IEEE Trans. Neural Sys. & Rehab. Eng., 17*, 312–321.

67. Ma, C. T., Mak, P. I., Vai, M. I., Mak, P., Pun, S. H., Feng, W., & Martins, R. P., (2009). A 90nm CMOS bio-potential signal readout front-end with improved powerline interference rejection. *IEEE Circ. and Sys.,* 665–668.

68. Wang, W. S., Wu, Z. C., Huang, H. T., & Luo, C. H., (2009). Low power instrumental amplifier for portable ECG. *IEEE Circ. and Sys. International Conference,* 1–4.

69. Gosselin, B., Sawan, M., & Chapman, C. A., (2007). A low-power integrated bioamplifier with active low-frequency suppression. *IEEE Trans. Biomed. Circ. & Sys., 1*, 184–192.

70. Harrison, R. R., & Charles, C., (2003). A low-power low-noise CMOS amplifier for neural recording applications. *IEEE J. Solid-State Circ., 38*(3), 958–965.

71. Wattanapanitch, W., Fee, M., & Sarpeshkar, R., (2007). An energy -efficient micropower neural recording amplifier. *IEEE Trans. Biomed. Circ. & Syst., 1*, 136–147.

72. Zhao, W., Li, H., & Zhang, Y., (2009). A low-noise integrated bioamplifier with active DC offset suppression. *IEEE Trans. Biomed. Circ. & Syst.,* 5–8.

73. Hu, Y., & Sawan, M., (2002). *CMOS Front-End Amplifier Dedicated to Monitor Very Low Amplitude Signal from Implantable Sensors* (Vol. 33, pp. 29–41). Kluwer Academic Publishers.

74. Popovic, D., Stein, B., Jovanovic, R. B., Dai, K. L., Kostovand, A. R., & Armstrong, W. W., (1993). Sensory nerve recording for closed-loop control to restore motor function. *IEEE Trans. Biomed. Eng., 40*(10), 1024–1031.

75. Stein, R., Charles, B., Davis, D., Jhamandas, L., Mannard, J. A., & Nichols, T. R., (1975). Principles underlying new methods for chronic neural recording. *Can. J. of Neuro. SCI.,* 235–244.

76. Hershenson, M., Boyd, S., & Lee, T., (2001). Optimal design of a CMOS op-amp via geometric programming. *IEEE Trans. Computer-Aided Design Integr. Circuits Sys., 20*(1), 1–21.

77. Hershenson, M., Boyd, S., & Lee, T. H., (1997). CMOS operational amplifier design and optimization via geometric programming. In: *Proc. 1st Int. Workshop Design Mixed-Mode Integrated Circuits Applicat.* (pp. 15–18). Cancun, Mexico.

78. Allen, P. E., & Holberg, D. R., (1987). *CMOS Analog Circuit Design* (1st edn.). U.K.: Oxford.

79. Ahuja, B. K., (1983). An improved frequency compensation technique for CMOS operational amplifiers. *IEEE J. Solid-State Circ., 18*, 629–633.

80. Yates, D., Lopez-Morillo, E., & Carvajal, R. G., (2007). A low-voltage low-power front-end for wearable EEG systems. *Conference of the IEEE EMBS.*

81. Huang, C. C., Hung, S. H., & Chung, J. F., (2008). Front-end amplifier of low-noise and tunable B/W gain for portable biomedical signal acquisition. *IEEE Circ. and Sys.,* 2717–2720.

82. Bronskowski, C., & Schroeder, D., (2006). An ultra-low-noise CMOS operational amplifier with programmable noise-power trade-off. *IEEE Solid-State Circ. Conf.,* 368–371.

83. Lentola, L., Mozzi, A., Neviani, A., & Baschirotto, A., (2009). *A 1uA Front-End for Pacemaker Atrial Sensing Channels.* IEEE.

CHAPTER 12

An Efficient Design of D Flip Flop in Quantum-Dot Cellular Automata (QCA) for Sequential Circuits

BIRINDERJIT SINGH KALYAN,[1] HARPREET KAUR,[2]
KHUSHBOO PACHORI,[2] and BALWINDER SINGH[3]

[1]*I. K. Gujral Punjab Technical University, University in Kapurthala, Punjab, India*

[2]*Department of Electronics and Communication Engineering, Guru Nanak Institute of Technical Campus Hyderabad, Telangana, India, E-mail: kaurharpr@gmail.com (H. Kaur)*

[3]*Center for Development of Advanced Computing (C-DAC), Mohali, Punjab, India*

ABSTRACT

Quantum cellular automata (QCA) is the most recent creating innovation for rapid, low power computation architectural technique in the field of new-age nanoelectronics that restructures the logical information as charge structures of as a Quantum cell which was firstly suggested by Lent et al. [1]. The Quantum-Dot cellular automata are an alternative for conventional CMOS technology to implement classic cellular automata with quantum dots. The use of quantum cell automata is an innovation as an option to CMOS technology on the nanoscale has a promising future as its integration with various digital circuits in a precisely low powered device. This chapter analyzes the QCA based flip-flops and proposed novel layouts of shift register and ring counter with less QCA cells and better performance parameters. The SISO shift register and ring counter is structured using a D flip flop which is redesigned using 38 cells, show 42% less complexity than previous structures.

12.1   INTRODUCTION

In this emerging technology era for the last two decades, the group of researchers at the University of Notre Dame was working on this promising technology paradigm shift transforms the nanotechnology research area. In their research, the architecture of QCA circuits which is a QCA cell designed from which the basic gates and logic devices are evolved. The research involved the physical level and then they designed the simulation tool QCA designer in 2004 [3]. Macucci [4] contributed in the manufacturing of QCA devices, and Lombardi et al. [5] presented the design and reliability test of digital circuits in QCA. The QCA technology has many advantages like its operation at low power and having high density operating at terahertz range [6–9]. The proposed designs of various flip flops can be effectively used to realize more complex memory circuits. The simulations of various Flip Flops in the present work have been carried out using QCA designer tool. The schematic Figure 12.1(a) is a QCA nanostructure having four quantum dots. The square placed the quantum dots at its corner. The cell comprises of two electrons that tunnel through the four quantum dots at the corner of the square cell. The tunneling did not take place between the two adjoining cells. The quantum dots (represented as *i)* in the cell where i=1, i=2, i=3, i=4 are the capacities that represents the interstitial position within the cell. Eqn. (12.1) represents the polarization in a cell.

$$P = \frac{(\rho_1 + \rho_3) - (\rho_2 + \rho_4)}{(\rho_1 + \rho_2 + \rho_3 + \rho_4)} \tag{12.1}$$

The polarization estimates the charge within the cell, up to which the charge is conveyed among the four dots. Logical representation is characterized in QCA in the forms of logiclogic.

Due to Coulombic attraction-repulsion [1] cells occupy the antipodal sites under the control clocking scheme in which fences between dots which is enough to free the electrons as shown in Figure 12.1(a). The two polarizations of −1 and +1 represent the binary "0" and "1," respectively. The propagation of the charges can be transmitted by the Coulombic repulsion within the cells within the cell array. There is interaction between the QCA cells which is highly irregular, indeed, even a somewhat spellbound in input cell actuated the enraptured output cell completely to its degree [10], if they are placed near to each other with the polarized cell. The binary information can be transferred by interacting with each other between adjusted cells onward inline of QCA cells termed as a "wire" as presented in Figure

12.1(b), where the quantum cells are neighboring each other instead of a physical wire. Interconnections between the different rationale parts are gotten to utilizing this QCA wire, ability to offer "processing-in-wire" [11]. QCA transferred the logical information without current flow and without the electrons tunneling between the two adjacent cells.

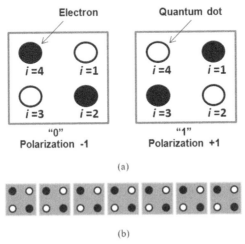

FIGURE 12.1   (a) Binary QCA cells; (b) QCA wire.
Source: Reprinted from Ref. [2]. Open access.

12.2   QCA MAJORITY IMPLEMENTATION

With the evolution of devices like parametrons and Esaki diodes in 1960 [12], the binary majority elements are also evolved with the development which describes how simple Boolean expressions can be implemented using a majority decision operator. On the basis of this approach, synthesis of complex networks of components which includes decomposition and rearrangement, transformed to majority element based which is having limited fan-in Ref. [13]. The synthesis take place with respect to various class of logic functions and Boolean expression. A distributive law showcases the majority logic is described in Refs. [15, 27]. The various algebraic method for logical designs, apply Veitch diagrams which is a geometric method for synthesis the various input majority gates n-argument switching functions [16] which is presented by Bernstein et al. [14]. The NOT Gate is represented by rearranging the QCA cells as shown in Figure 12.2 and majority gate, respectively. By applying input "Logic 0," we get output of "Logic 1" which amounts to inverting the input or vice-versa. The output of the Three input MG shown in Figure 12.2(b) having three inputs. If X, Y, and Z are three inputs to the MG, then the output is given by: $M(X,Y,Z) = XY + YZ + XZ$.

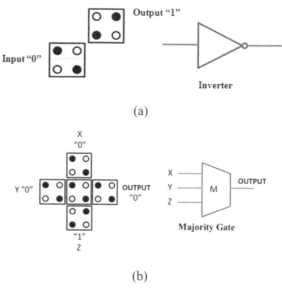

(a)

(b)

FIGURE 12.2   (a) Inverter; (b) three input majority gate $M(X, Y, Z) = XY + YZ + AZ$.
Source: Reprinted from Ref. [2]. Open access.

If $Z = 0$, the three input MG will be structured as an AND gate. In other words, $M(X,Y, 0) = XY$. Likewise, by keeping $Z = 1$, the MG functions will act as an OR gate $M(X, Y, 1) = (X + Y)$. A MG technique has advantages compared to an AND or OR semiconductor-based gate [20–23] as they require just a similar number of cells in QCA then that of other gate fabricated technologies where the 3-input MG which is less than required then the number of transistors for some other gates in applying the gates through majority gate [24, 25].

Figure 12.3 represents the five input MG structure which is designed for complex QCA circuit implementations [17]. This will reduce the hardware requirements for more complex and simply the five input QCA logic circuits [18]. The majority structure uses the 10 QCA cells for the implementation where A, B, C, D, E are the input logic of the five-input MG structure. The logical declaration for the output of five-input MG is represented by Eqn. (12.2). By using this equation designer can implement more complex logic. A, B, C, D are referred as X_1, X_2, X_3, X_4, X_5

$$M(X_1, X_2, X_3, X_4, X_5) = X_1X_2X_3 + X_1X_2X_4 + X_1X_2X_5 + X_1X_3X_4 + X_1X_3X_5 +$$
$$X_1X_4X_5 + X_2X_3X_4 + X_2X_3X_5 + X_2X_4X_5 + X_3X_4X_5 \qquad (12.2)$$

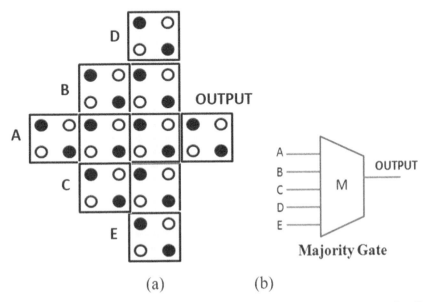

(a)   (b)

FIGURE 12.3 (a) QCA based majority gate; (b) schematic of majority gate having five inputs.
Source: Reprinted from Ref. [2]. Open access.

12.3 QUANTUM-DOT CELLULAR AUTOMATA BASED FLIP FLOP DESIGN

The Flip Flop circuit is a major component for the design of any memory device. This is two output logics, one with the normal output and other stored bit is complement of the first. The Binary information fed to various flip-flop which gives rise to different logical outputs which use to implement in various memory devices. The types of flip flop under consideration:

- JK FF;
- T FF;
- SR FF;
- D FF.

For minimizing the complexity of various flip flops design, majority logic is employed [19]. To keep the memory in motion and intact, there is a need to create a feedback loop in QCA therefore proposed structures of the flip-flops are simulated in this chapter. For the generation of QCA architecture of various flip flops, the characteristics equation is generated from the existing truth tables of these flip flops.

12.3.1 JK FLIP FLOP

Simultaneously when logic 1 is applied to J and K, the FF transforms to its supplement state. The JK FF vectors are established in Table 12.1. A sequential JK flip-flop is shown in Figure 12.4. J and K are the two inputs along with one control input clock (clk). The JK flip flop is having two inverters and four majority gates as demonstrated in Figure 12.5.

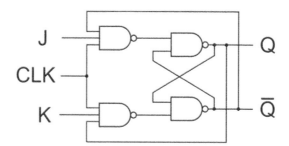

FIGURE 12.4   Logic design of JK FF.
Source: Reprinted from Ref. [2]. Open access.

TABLE 12.1   Simulating Vectors of JK FF

Inputs		Active	0	1	2	3	4
▦	J	☑	☑	☐	☑	☐	☑
▦	K	☑	☐	☑	☑	☐	☐
▦	Clock	☑	☑	☐	☑	☐	☑

Source: Reprinted from Ref. [2]. Open access.

$$Q(t+1) = JQ' + K'Q \qquad (12.3)$$

The majority gate M1 is feed with output Q and majority gate M3 is feed with the complement of the output from the flip flop. The output from M3 and M1 are synchronized with the clk input by majority gate M2. Thereby another majority gate M2 act as an OR gate to produce the output JQ'+K'Q which is act as input to gate M4 which is utilized and derived in Eqn. (12.3).

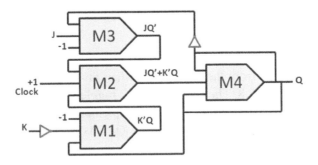

FIGURE 12.5   Majority gate based JK FF [2].

The layout of JK FF is implemented in QCA Designer shown in Figure 12.6. The QCA accomplishment with 46 cells to design, with an area of 38,804 nm^2 (Figure 12.7).

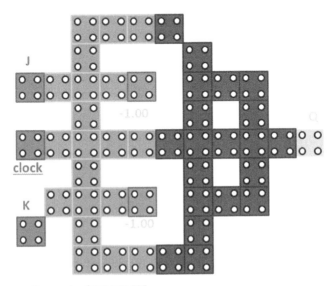

FIGURE 12.6   Layout of JK FF [2].

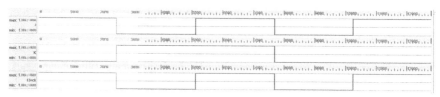

FIGURE 12.7   Simulation results of JK FF.

12.3.2   T FLIP FLOP

The T FF is derived from the JK type FF when inverter connects the inputs as shown in Figure 12.8. The state of T FF is toggled during the transaction. When clock pulse is triggered, flip flop complements the present state. The simulation vector Table 12.2 which represents its toggling switching into various states.

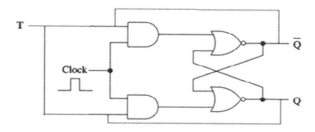

FIGURE 12.8   Logic design of T FF.

TABLE 12.2   Simulating Vectors of T FF

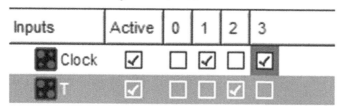

Inputs	Active	0	1	2	3
▦ Clock	☑	☐	☑	☐	☑
▦ T	☑	☐	☐	☑	☐

$$Q(t+1) = TQ' + T'Q \tag{12.4}$$

The structure of T FF is structured by implementing 4 MGs and 2 inverters presented in Figure 12.9. The input of majority gate M4 is having three inputs, first which is complement from Majority gate M3, second is the feedback from output, last is the output from majority gate M2. On the other hand, M1 and M3 act as OR gate to produce TQ'+T'Q under the influence of clock input. The desired characteristics are produced by MG M4 which form a loop for the Eqn. (12.4).

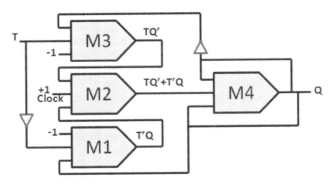

FIGURE 12.9   Majority gates-based T FF.

The T flip flop is executed in QCA designer tool of the T flip flop is shown in Figure 12.10 with 53 cells and occupied the area of 49,484 nm². The simulation results are produced in Figure 12.11.

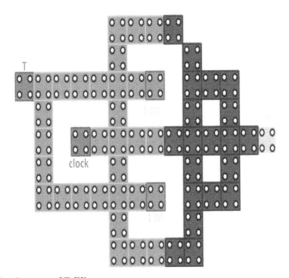

FIGURE 12.10   Layout of T-FF.

FIGURE 12.11   Simulation results of T flip flop.

12.3.3   SR FLIP FLOP

The SR flip-flop is enabled by a shown in Figure 12.12, contribution by two NOR gate and two AND gates. As long as the clock pulses remain zero, the output remains at zero in spite of the S and R input values.

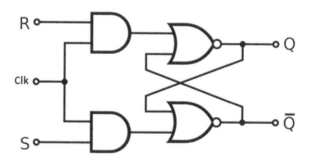

FIGURE 12.12   Schematic of SR FF.

The simulating vector of SR FF is shown in Table 12.3. This Eqn. (12.5) indicates the estimation of the following state as an element of the present state with the implementation of clock pulse is derived by using K-Map. The SR flip flop is reproduced by utilizing five majority gates and one inverter, as shown in Figure 12.13. S and R are the two inputs of the circuit with one control input clock. The output of the flip flop is feedback to M4 acts as an AND gate which combines with the input R' of majority gate M3 and output S+R'Q to produce R'Q. The loop is formed by the M4 and M5 majority gates. The OR operation is designated by M5 with the combination of "S" of the output of majority gate M1 and "R'Q" of majority gate M4. The desired symptomatic Eqn. (12.5) of SR FF at the output of M5.

TABLE 12.3 Simulating Vector of SR Flip Flop

Inputs	Active	0	1	2	3	4	5	6
S	✓	✓	✓	✓	☐	☐	☐	☐
R	✓	☐	✓	☐	✓	✓	☐	☐
clock	✓	✓	☐	✓	☐	✓	☐	✓

$$Q(t+1) = S + R'Q \tag{12.5}$$

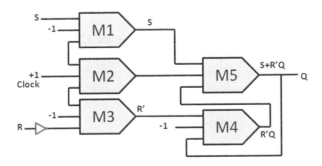

FIGURE 12.13 Majority gate-based SR FF.

The layout of SR FF is implemented using QCA cells having five majority gates as shown in Figure 12.14 with 64 cells and area of 60,888 nm². The simulation result of QCA based SR flip flop is shown in Figure 12.15.

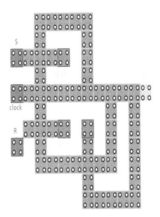

FIGURE 12.14 QCA based SR FF.

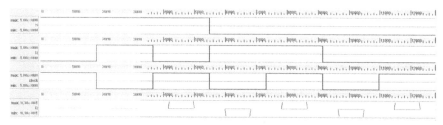

FIGURE 12.15   Simulation results of SR flip flop.

12.3.4   D FLIP FLOP

The logic schematic of D FF presented in Figure 12.16. The D FF is a reformation of SR FF which is enabled by clock signal. The NOT gate is installed in between the two SR FF inputs with clock signal. The D acts as input with a clock signal and the complement of the D input goes to the gate input.

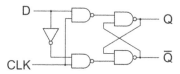

FIGURE 12.16   Logic schematic of D flip flop.

When clock pulse is triggered, logical information available at the input of flip flop which is transferred to the output 'Q.' The simulating vector table of the D FF is shown in Table 12.4. The D FF circuit is controlled by a clock along with one input D. The structure of D FF is constructed with three input majority gate and one inverter with the transit characteristic of Eqn. (12.4) as shown in Figure 12.17. The majority gate M1 simulate as an AND gate and majority gate M2 simulate as an OR gate. The output of the majority gate M1 and M2 triggered with the majority gate M3 with feedback from the output simulates the characteristics Eqn. (12.6).

TABLE 12.4   Simulating Vector of D Flip Flop

Inputs	Active	0	1	2	3	4	5
clock	✓	✓	☐	✓	☐	✓	☐
D	✓	☐	☐	☐	✓	✓	✓

$$Q(t+1) = D \qquad (12.6)$$

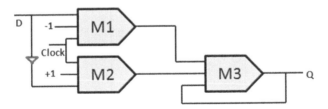

FIGURE 12.17   Majority gate-based D FF.

The D FF is implemented using QCA designer presented in Figure 12.18 with 38 cells and an area of 47,838 nm^2 and having 42% lesser complexity than the previous structure. The simulation results are shown in Figure 12.19.

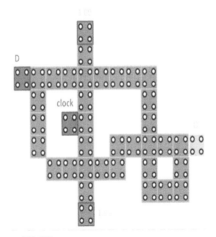

FIGURE 12.18   Layout of D-FF.

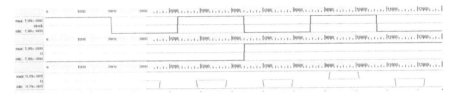

FIGURE 12.19   Simulation results of D FF.

The performance analysis proposed JK FF, T FF, SR FF and D FF are given in Table 12.5. The performance analysis of proposed flip flops circuits is compared in context to their complexity, area, and latency.

TABLE 12.5 Performance Analysis of Flip Flops

Flip Flop	Previous Structure			Proposed Structure			
	Complexity	Area (nm)	Total Area (nm²)	Complexity	Area (nm)	Total Area (μm²)	Latency	Quantum Cost (Latency)² × Area
JK	50 [34]	198 × 318	62,964	46	218 × 178	0.038	0.5	0.01
T	66 [35]	338 × 178	60,164	53	278 × 178	0.049	0.5	0.01
SR	66 [36]	418 × 218	91,124	59	236 × 218	0.051	1	0.05
D	48 [33]	218 × 218	47,524	38	238 × 201	0.047	1	0.04

12.4   SHIFT REGISTER USING D-FF

The serial in/serial-out shift register accepts and produces the stored data serially, one bit at a time on a single line as shown in Figure 12.20. The clock pulse is provided from the common source and the same instant. The shift register is constructed from single DFF which is designed in Figure 12.18 with 38 cells having the total area of 47,838 nm^2, now the shift register serial in serial out is having the 4 D-FF, total cell used are 211 with the area spacing of 1,058 nm × 241 nm that is 254,978 nm^2. The layout design of shift registers by QCA designer tool in Figure 12.21 having four D-FF are arranged serially to form the 4-Bit SISO register. The input is fed through input D and the clock is provided as shown in Table 12.6 as the design vectors of shift register and the simulation waveform in Figure 12.22 is thus formed using these vectors using the Euler method in coherence vectors simulation engine as shown in Table 12.7.

TABLE 12.6   Design Vector

Inputs	Active	0	1	0	1
D	☑	☑	☐	☑	☐
clock	☑	☐	☑	☐	☑

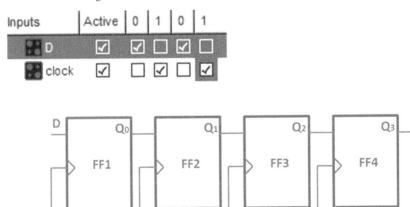

FIGURE 12.20   Block diagram of shift register using DFF.

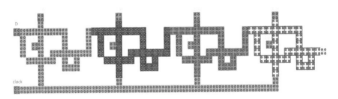

FIGURE 12.21   Layout of the serial-in serial-out shift register using DFF.

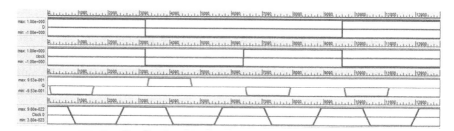

FIGURE 12.22   Simulation waveform of serial-in serial-out shift register.

12.5   RING COUNTER USING D-FF

The ring counter is designed using 38 cell D FF, the design is ring counter schematic as presented in Figure 12.23. The structure of counter has 272 cells having a total area of 1,083 nm × 258 nm means 279,414 nm². The feedback from the output is fed back to the FF1 from FF4. The clock pulse is given at the same instant; all four D-Flip Flop is having different clock pulse. The basic D-Flip Flop is designed having 38 cells in Figure 12.18. The layout is designed with four D flip flop which are serially designed with feedback as shown in Figure 12.24. The design vector is shown in Table 12.6 which formed the simulation waveform is described in Figure 12.25 using Euler method in coherence vectors simulation engine as shown in Table 12.7. The design vectors are the input provided for simulating the design in given truth tabular format.

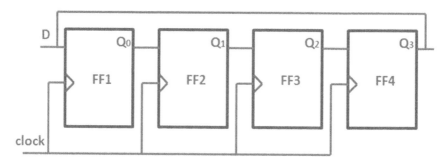

FIGURE 12.23   Block diagram of shift register using DFF.

TABLE 12.7 Coherence Vector Simulation Engine

Coherence Vector Optio...	— □ ×	
Temperature:	1.000000	K
Relaxation Time:	1.000000e-015	s
Time Step:	1.000000e-016	s
Total Simulation Time:	7.000000e-011	s
Clock High:	9.800000e-022	J
Clock Low:	3.800000e-023	J
Clock Shift:	0.000000e+000	
Clock Amplitude Factor:	2.000000	
Radius of Effect:	80.000000	nm
Relative Permittivity:	12.900000	
Layer Separation:	11.500000	nm
⊙ Euler Method		
○ Runge Kutta		
☑ Randomize Simulation Order		
☐ Animate		
✗ Cancel	✔ OK	

There are 12,800 samples with convergence tolerance 0.001000, radius of effect of cell is 65 nm and maximum iterations per sample is 100 in bistable simulations in the simulating engine of QCA Designer tool (Table 12.8) [26].

TABLE 12.8 Performance Evaluations

Layout	No of QCA Cells	Area	Total Area
Shift register	211	1,058 nm × 241 nm	254,978 nm²
Ring counter	272	1,083 nm × 258 nm	279,414 nm²

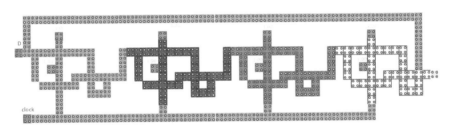

FIGURE 12.24 Layout of ring counter using DFF.

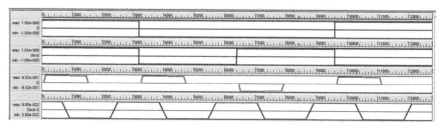

FIGURE 12.25   Simulation waveform for ring counter using D-flip flop.

12.6   CONCLUSION

The QCA layout of various FF, Shift Register, Ring Counter is created from QCA Designer tool. QCA Designer Pro tool is used to estimate the power in the circuits [29–32], which provide pliability in terms of spacing and dimensions of QCA cells in a layout in the presence of radiations. The size depends upon the number of QCA cells utilized to design that structure, hence it is possible to minimize the area by reducing the number of cells which is only possible by using majority gates. The D-FF Flip Flop found ideal for the design of sequential circuits like 4-bit shift register and ring counter having the latency of 1 in both designs. The shift register and ring counter are novel designs having complexity much lesser with the use of D flip flop whose complexity is of 38 cells and area of 0.047 μm^2. The shift register is designed with complexity of 211 cells and latency of 1 similarly ring counter with complexity of 272.

KEYWORDS

- **clock**
- **flip flop**
- **majority logic**
- **quantum cellular automata**
- **ring counter**
- **serial in serial out**
- **shift register**

REFERENCES

1. Craig, S. L., Douglas, T. P., Wolfgang, P., & Gary, H. B., (1993). Quantum cellular automata. *Nanotechnology, 4,* 49–57.
2. Kalyan, B. S., & Singh, B., (2018). Quantum-dot cellular automata (QCA) based 4-bit shift register using efficient JK flip flop. *International Journal of Pure and Applied Mathematics, 118*(19), 143–157.
3. Walus, K., Dysart, T. J., Jullien, G. A., & Budiman, R. A., (2004). QCA designer: A rapid design and simulation tool for quantum-dot cellular automata. *IEEE Transactions on Nanotechnology, 3,* pp. 26–31.
4. Massimo, M., (2006). *Quantum Cellular Automata: Theory, Experimentation and Prospects.* Imperial College Press, London.
5. Fabrizio, L., & Jing, H., (2007). *Design and Test of Digital Circuits by Quantum-Dot Cellular Automata.* Norwood, MA: Artech House, Inc.
6. Yuhui, L., Mo, L., & Craig, L., (2007). Molecular quantum-dot cellular automata: From molecular structure to circuit dynamics. *Journal Appl. Phys., 102*(3), 1–7.
7. Yuhui, L., Mo, L., & Lent, C., (2006). Molecular electronics—from structure to circuit dynamics. In: *Proc. 6th IEEE Conf. Nanotechnology* (pp. 62–65).
8. Enrique, P. B., Eric, Y., & Craig, S. L., (2010). Power dissipation in clocking wires for clocked molecular quantum-dot cellular automata. *Journal Computational Electronics, 9*(1), 49–55.
9. Sarah, F., Arun, R., Andrew, J., Randal, R., & Peter, K., (2002). Memory in motion: A study of storage structures in QCA. In: *Proc. 1st Workshop Non-Silicon Computing,* 1–8.
10. Douglas, T. P., & Craig, S. L., (1994). Logical devices implemented using quantum cellular automata. *Journal of Applied Physics, 75,* 1818–1825. doi: 10.1063/1.356375.
11. Cohn, M., & Lindaman, R., (1961). Axiomatic majority-decision logic *IRE Transactions on Electronic Computers, EC-10*(1).
12. Yuhui, L., & Craig, S. L., (2005). Theoretical study of molecular quantum-dot cellular automata. *Journal of Computational Electronics, 4*(1), 115–118.
13. Fusachika, M., (1963). Realization of arbitrary logical functions using majority elements. *IEEE Trans. Electronic Computers, EC-12*(3), 183–191.
14. Haider, et al., (2009). Controlled coupling and occupation of silicon atomic quantum dots at room temperature, *Physical Review Letters, 102*(4), 805–846. doi: 10.1103/PhysRevLett.102.046805.
15. Sheldon, B. A., (1961). On the algebraic manipulation of majority logic. *IEEE Trans. Electronic Comput. EC, 10*(4), 779–779.
16. Miller, H. S., & Winder, R. O., (1962). Majority logic synthesis by geometric methods. *IEEE Trans. Electronic Computers, EC-11*(1), 89–90.
17. Keivan, N., Razieh, F., Samira, S., & Mostafa, R. A., (2010). A new quantum-dot cellular automata full-adder. *Microelectronic Journal,* 820–826.
18. Navi, K., Sayedsalehi, S., Farazkish, R., & Azghadi, M. R., (2010). Five-input majority gate, a new device for quantum-dot cellular automata. *J. Computer Theory Nanoscience, 7*(8), 1546–1553.
19. Heumpil, C., & Earl, E. S., (2007). Adder designs and analyses for quantum-dot cellular automata. *IEEE Transactions on Nanotechnology,* 374–383.

20. Gardelis, S., Smith, C. G., Cooper, J., Ritchie, D. A., Linfield, E. H., & Jin, Y., (2002). Evidence for transfer of polarization in a quantum-dot cellular automata cell consisting of semiconductor quantum dots. *Physical Review B, 67*(3), 1–3.

21. Perez-Martinez, F., Farrer, I., Anderson, D., Jones, G. A. C. Ritchie, D. A., Chorley, S. J., & Smith, C. G., (2007). Demonstration of a quantum cellular automata cell in a GaAs/AlGaAs heterostructures. *Applied Physics Letter*, 1–3.

22. Smitha, C. G., Gardelisa, S., Rushfortha, A. W., Crooka, R., Coopera, J., Ritchiea, D. A., Linfielda, E. H., et al., (2003). Realization of quantum-dot cellular automata using semiconductor quantum dots. *Superlattices and Microstructures*, 195–203.

23. Mitic, M., Cassidy, M. C., Petersson, K. D., Starrett, R. P., Gauja, E., Brenner, R., Clark, R. G., et al., (2006). Demonstration of a silicon-based quantum cellular automata cell. *Applied Physics Letters*, 1–3.

24. Lent, C. S., & Tougaw, P. D., (1997). A device architecture for computing with quantum dots. In: *Proceeding of the IEEE* (Vol. 85, No. 4, pp. 541–557).

25. Yamaguchi, K., (1983). A mobility model for carriers in the MOS inversion layer. *IEEE Trans. Electron Devices, 30*, 658–663.

26. QCA Designer Tool Version 2.0.3. https://download.freedownloadmanager.org/Windows-PC/QCADesigner/FREE.html (accessed on 21 September 2021).

27. Bernsteina, G. H., Imrea, A., Metlushkoc, V., Orlova, A., Zhoua, L., Jia, L., Csabab, G., & Poroda, W., (2005). Magnetic QCA systems. *Microelectronic Journal, 36*, 619–624,

28. Craig. S. L., Douglas, T., & Wolfgang, P., (1994). Quantum cellular automata: The physics of computing with arrays of quantum dot molecules. In: *Proceedings of the Workshop on Physics and Computing*, 5–13.

29. Bonci, L., & Macucci, M., (2006). Analysis of power dissipation in clocked quantum cellular automaton circuits. *Solid-State Device Research Conference ESSDERC*, 57–60.

30. Srivastava, S., Sarkar, S., & Bhanja, S., (2006). Power dissipation bounds and models for quantum-dot cellular automata circuits. *IEEE Conference on Nanotechnology, 1*, 375–378.

31. Bandyopadhyay, S., (2007). Power dissipation in spintronic devices: A general perspective. *Journal of Nanoscience and Nanotechnology, 7*, 168–180.

32. Mikhail, B. M., (2003). Dissipation and decoherence in quantum systems. *Physics-Uspekhi, 46*(11), 1163–1182.

33. Sara, H., & Keivan, N., (2012). New robust QCA D flip flop and memory structures. *Microelectronics Journal*, 929–940. doi: 10.1016/j.mejo.2012.10.007.

34. Kun, K., Yun, S., & Ruqian, L., (2010). Counter designs in quantum-dot cellular automata. 10th *IEEE International Conference*. doi: 10.1109/NANO.2010.5698033.

35. Mohammad, T., (2011). A new architecture for T flip flop using quantum-dot cellular automata. *IEEE 3rd Asia Symposium on Quality Electronic Design*. doi: 10.1109/ASQED.2011.6111764.

36. Huang, J., Momenzadeh, M., & Lombardi, F., (2007). Design of sequential circuits by quantum-dot cellular automata. *Microelectr. J., 38*, 525–537. doi: 10.1016/j.mejo.2007.03.013.

CHAPTER 13

Design and Performance Analysis of Digitally Controlled DC-DC Converter

SUBHRANSU PADHEE,[1] MADHUSMITA MOHANTY,[2] and AMBARISH PANDA[3]

[1]Department of Electrical and Electronic Engineering,
Aditya Engineering College, Surampalem, Andhra Pradesh, India,
E-mail: subhransupadhee@gmail.com

[2]Department of Electrical and Electronics Engineering,
National Institute of Technology, Puducherry, Karaikal, India

[3]Department of Electrical and Electronics Engineering, Sambalpur
University Institute of Information Technology, Burla, Odisha, India

ABSTRACT

This work provides a systematic overview of the design and performance of voltage-mode digital control of DC-DC converter. Different design aspects of digital control have been highlighted and digital implementation of PID controller and pulse width modulator scheme has been discussed. A case study of DC-DC buck converter is used to illustrate the working of digital controller.

13.1 INTRODUCTION

The design of power electronic system for different applications is a typical engineering problem where the design professional has to consider cost, performance, system complexity, efficiency, reliability, and robustness. As the power system is becoming increasingly complex in nature, power management issues in the system are becoming a real challenge. The complex system comprises of multiple subsystems which have either full or partial interaction with other subsystems. Therefore, power management issue has been one of the emerging research topics in recent times. Besides the control issue, modern day system has several additional features depending on the application. Some of the additional features are (a) communication, (b) reprogramming, (c) automatic tuning, (d) fault diagnosis, etc.

Classical technique to design a compensator for DC-DC converter relies on the well-established technique of frequency response method where compensator is designed via the frequency response of the system. Using this technique, different controller is designed which can be implemented using operational amplifier-based circuits. The analog controller has a large bandwidth, but its ineffective when there are changes in the dynamics of the system. The lack of reprogramming and updation of controller parameters makes the controller fixed in nature (non-adaptive). The aging of the components and variation of load also plays a vital role where analog controller fails to provide adequate control output. Analog controller has a large number of parts which increases the design complexity and the computational complexity of the controller is also poor.

Therefore, digital controller has been used by power electronic design professional to overcome the limitations of the analog controller. Digital controller is reprogrammable and the controller parameters can be tuned according to the requirements. Digital controller possesses good computational complexity which allows the designer to provide additional features in the controller. The introduction of digital signal processor in 1983 by Texas Instrument paved the way of actual digital control in DC-DC converter [1, 2]. Figure 13.1 provides a system-level architecture of a digitally controlled DC-DC converter which is a combination of analog as well as digital systems. DC-DC converter, gate driver, sensor, and signal conditioning circuit comprise of the analog system whereas digital compensator, digital pulse width modulator constitutes the digital part.

Design of power electronic system for different applications is a typical engineering problem where the design professional has to consider cost, performance, system complexity, efficiency, reliability, and robustness. As the power system is becoming increasingly complex in nature, power management issues in the system are becoming a real challenge. The complex system comprises of multiple subsystems which have either full or partial interaction with other subsystems. Therefore, power management issue has been one of the emerging research topics in recent times. Beside the control issue, modern day system has several additional features depending on the application. Some of the additional features are (a) communication; (b) re-programming; (c) automatic tuning; (d) fault diagnosis, etc.

Classical technique to design a compensator for DC-DC converter relies on the well-established technique of frequency response method where compensator is designed via the frequency response of the system. Using this technique, different controller is designed which can be implemented using operational amplifier-based circuits. The analog controller has a large bandwidth, but its ineffective when there are changes in the dynamics of the system. The lack of reprogramming and updation of controller parameters makes the controller fixed in nature (non-adaptive). The aging of the components and variation of load also plays a vital role where analog controller fails to provide adequate control output. Analog controller has a large number of parts which increases the design complexity and the computational complexity of the controller is also poor.

Therefore, digital controller has been used by power electronic design professional to overcome the limitations of the analog controller. Digital controller is reprogrammable and the controller parameters can be tuned according to the requirements. Digital controller possesses good computational complexity which allows the designer to provide additional features in the controller. The introduction of digital signal processor in 1983 by Texas Instrument paved the way of actual digital control in DC-DC converter [1, 2].

Figure 13.1 provides a system-level architecture of a digitally controlled DC-DC converter which is a combination of analog as well as digital systems. DC-DC converter, gate driver, sensor, and signal conditioning circuit comprise of the analog system whereas digital compensator, digital pulse width modulator constitutes the digital part.

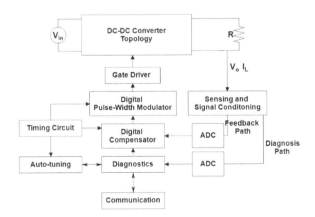

FIGURE 13.1 System-level architecture of digitally controlled DC-DC converter.

DC-DC converter provides a regulated output with the help of a feedback controller and a modulator unit. For digital control scheme, the sensors (voltage and current sensor) and signal conditioning circuits are employed to measure the instantaneous voltage and current of the converter. The measured value is converted to discrete format using analog to digital converter (ADC). ADC operating in uniform sampling is considered in this study. To select an appropriate ADC for a specific application, ADC resolution n_{ADC} and conversion time of ADC Δt_{ADC} is considered.

The digital compensator takes the measured value as input and provides a control signal as output. The output of the digital compensator is provided to digital pulse width modulation (DPWM) unit which provides adequate gating pulse to the DC-DC converter. The resolution of DPWM should be much higher than ADC to avoid undesirable quantization effects such as limit cycle oscillation. Gate driver circuit provides the high current turn-on/turn-off driving commands to the power switches (MOSFET or IGBT) by performing necessary voltage conversion operation. Gate driver circuit provides the necessary turn-on and turn-off command with a specific delay time Δt_g. Modern ASIC based gate driver circuit has a typical delay time of 50 ns. With the advent of newer technology, additional safety, diagnosis, communication, and self-tuning features have been included in the system. Safety and diagnosis system are essential components in high-reliability applications such as server, data center, aircraft, spacecraft, and military applications. Communication link allows the system to communicate with systems of the outside world and it is also useful to receive as well as send information and command from outside world. Autotuning feature allows the digital compensator to reprogram the control law automatically when the need arises. Using digital control scheme, system identification and health monitoring of DC-DC converter can also be performed [3]. Digital controller design technique for different switched-mode power supply has been discussed in Ref. [4]. Digital control of synchronous buck converter has been illustrated in Ref. [5]. Direct-digital approach for digital controller of power converter has been discussed in Ref. [6]. Look-up-table (LUT) based DPWM scheme implementation for SoC implementation has been discussed in Ref. [7]. Multi sampled digitally controlled DC-DC converter has been discussed in Ref. [8], where ripple-elimination is carried out using multi-sampled approach. Voltage and current mode control of buck converter using FPGA has been discussed in Ref. [9]. Limit cycle oscillation reduction techniques has been proposed in Ref. [10]. PID controller using FPGA has been discussed in Ref. [11]. Programmable controller IC for DC-DC converter in power factor correction application has been discussed in Ref. [12]. The digital controller can be implemented using different technologies such as:

- Microcontroller;
- Digital signal processor (DSPs);
- Field programmable gate array (FPGA) and complex programmable logic device (CPLD);
- Industrial computers (PXI and VXI bus system).

Comparative analysis of different computational platform has been studied in Ref. [13]. Among different computational platforms, FPGA is reconfigurable in nature, has a parallel computational structure and has a speed which is many a times higher than DSPs. FPGA provides fixed-point format and is treated as an ideal case of developing digital controller for switched-mode power converter. The word-length issue of other computational platform has been removed in FPGA as it has flexible precision arithmetic [14–16]. The difference between FPGA and ASIC has been studied in Ref. [17]. Comparative review of DSPs and FPGA based controller implementation for DC-DC converter has been studied in Ref. [18]. Apart from above platforms, FPGA based multi-processor system-on-chip (MPSoC) has been used for multi-thread processing in embedded control [19]. FPGA based controllers are widely used in power electronics and electrical machine drive applications [20, 21]. Digital controller for dynamic voltage scaling (DVS) processor enabled DC-DC converter [22]. A digital controller with four functionalities for DC-DC converter has been discussed in Ref. [23].

Despite the advantages of digital control techniques, there are some inherent limitations. Though the higher switching frequency of DC-DC converter reduces the size and rating of passive components, it also causes electromagnetic interference (EMI) problems. One of the main methods to limit the EMI is to use frequency-modulated pulse width modulation (PWM) scheme. The mathematical analysis of the spectrum of frequency-modulated PWM signal has been discussed in Ref. [24].

13.2 DIGITAL VOLTAGE MODE CONTROL OF DC-DC CONVERTER

Figure 13.2 provides the voltage mode digital control scheme of switching power converter. Switching power converter comprises of input DC supply, controlled switch, uncontrolled switch, passive filters (*L* and *C*) and load resistance. The switching power converter is considered a DC-DC converter.

In this study, DC-DC buck converter is considered as the switching power converter.

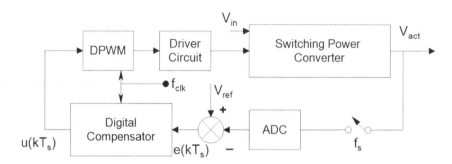

FIGURE 13.2   Voltage mode digital control scheme of switching power converter.

The output voltage across the load resistance of the DC-DC converter is $V_{act}(t)$. The output voltage is passed through ADC with sampling period T_s to generate a sampled output voltage $V_{act}(kT_s)$. The resolution of ADC is represented as:

$$N_{adc} = \text{int}\left[\log_2\left(\frac{V_{\text{max\_adc}}}{V_{ref}}\right)\right]\left(\frac{V_o}{\Delta V_o}\right) \tag{13.1}$$

where; V_0 is the output voltage; and ΔV_0 is the change in output voltage.

There is different type of ADC which can be classified according to resolution, speed, and noise immunity. Table 13.1 provides a comparative performance of ADC.

TABLE 13.1   Comparative Performance of ADC

Design	Speed	Resolution	Noise Immunity
Successive approximation	Medium	10–16 bits	Poor
Integrating	Slow	12–18 bits	Good
Ramp/counting	Slow	14–24 bits	Good
Flash/parallel	Fast	4–8 bits	None

The reference voltage $V_{ref}(t)$ is compared with the actual voltage $V_{act}(t)$ and a resultant error signal is generated $e(kT_s)$. The error signal is represented in signed integer (N bits) data type. The error signal will assume the value of maximum error $\left(\dfrac{+2^N}{2}\right)$, when it is greater than desired maximum level. The error signal will assume the value of minimum error $\left(\dfrac{-2^N}{2}\right)$, when it is lower than desired minimum level. The error signal is passed through the digital compensator to produce the new value of controlled output $u(kT_s)$. The controller output is supplied to DPWM which generates the essential PWM signal which drives the switch of the converter via a gate driver. A detailed analysis of the voltage mode digital control scheme is discussed in further sections.

13.3   DESIGN AND MODELING OF DC-DC BUCK CONVERTER

For voltage-mode digital control scheme of DC-DC buck converter, design, and modeling of the buck converter is essential. Figure 13.3 provides the circuit diagram of DC-DC buck converter where the input voltage is V_i, inductor L, parasitic resistance of inductor r_L, capacitor C, equivalent series resistance of capacitor r_c, controlled switch S and diode D. The output voltage is V_0 and load resistance is R_L.

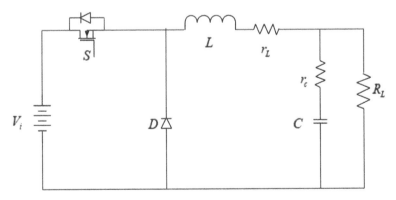

FIGURE 13.3   Circuit diagram of DC-DC buck converter.

The following assumptions have been considered during the design and modeling of DC-DC buck converter.

$$v_o(t) = V_o + V_{ripple}(t) \tag{13.2}$$

In a well-designed converter, the voltage ripple V_{ripple} is required to be less than 2% of the DC component under steady-state condition.

$$\left\| V_{ripple}(t) \right\| \lll V_o \qquad (13.3)$$

Using the approximation in Eqn. (13.3) in Eqn. (13.2), we get:

$$V_o(t) \approx V_O \qquad (13.4)$$

For steady-state operation of an inductor in a DC-DC converter, net inductor voltage in a switching period must be zero, i.e., $\langle V_L \rangle_T = 0$

For steady-state operation of a capacitor in a DC-DC converter, net capacitor current in a switching period must be zero, i.e., $\langle i_c \rangle_T = 0$

When switch is ON:

$$\Delta I_L = I_{L(\max)} - I_{L(\min)} = \frac{V_i - V_o}{L} DT_s \qquad (13.5)$$

When switch is OFF:

$$\Delta I_L = I_{L(\max)} - I_{L(\min)} = \frac{V_o}{L}(1-D)T_s \qquad (13.6)$$

The minimum value of inductor required to keep the DC-DC converter in continuous conduction mode (CCM) is given by:

$$L_{\min} = \frac{1-D}{2f_s} R_L \qquad (13.7)$$

where; f_s is the switching frequency of the converter.

The capacitor voltage ripple of the converter is given as:

$$\frac{\Delta V_o}{V_o} = \frac{1-D}{8LCf_s^2} \qquad (13.8)$$

For small-signal modeling of the DC-DC buck converter, the state-space averaging method is applied considering the converter is operating in CCM. Considering inductor current and capacitor voltage as independent state-variables and duty ratio of the converter as averaging parameter, the state-space model of the converter is formulated. The transfer function of DC-DC buck converter considering all parasitic can be represented as:

$$G_{dv}(s) = \frac{V_i(Cr_c s + 1)}{s^2 LC\left(\dfrac{r_c + R_L}{r_L + R_L}\right) + s\left(r_c C + C\left(\dfrac{r_L R_L}{r_L + R_L}\right) + \dfrac{L}{r_L + R_L}\right) + 1} \qquad (13.9)$$

The generalized form of control to output voltage of DC-DC converter can be represented as:

$$G_{dv}(s) = G_o \frac{1 + \dfrac{s}{\omega_{zesr}}}{1 + \dfrac{s}{Q\omega_o} + \left(\dfrac{s}{Q\omega_o}\right)^2} \qquad (13.10)$$

where; ω_0 is the corner frequency of buck converter $\omega_o = \sqrt{\dfrac{R_L + r_L}{LC(R_L + r_c)}}$, Q is

the quality factor $Q = \dfrac{1}{\omega_o\left(r_c C + \dfrac{L}{r_L + R_L} + \dfrac{R_L r_L C}{R_L + r_L}\right)}$, is DC gain $G_o = \dfrac{V_o}{D}$, ω_{zesr} is the

zero-frequency corresponding to equivalent series resistance $\omega_{zesr} = \dfrac{1}{Cr_c}$

Converting the continuous transfer function to discrete domain, the discrete domain transfer function can be represented as:

$$G_{dv,z} = \frac{b_1 z^{-1} + b_2 z^{-2}}{1 + a_1 z^{-1} + a_2 z^{-2}} \qquad (13.11)$$

13.4   DIGITAL COMPENSATOR

Let us consider a non-interacting type PID controller represented as:

$$G_c(s) = K_p + \frac{K_i}{s} + K_d s \qquad (13.12)$$

The non-interacting type PID controller can be implemented using parallel form where three-gain terms can be tuned according to a specific design criterion. The analog form of PID controller can be implemented using operational amplifier circuits along with RC networks.

Discrete-time PID controller can be derived by discretizing the analog counter part of the PID controller. Backward Euler discretization method $\left(s = \dfrac{1-z^{-1}}{T} \right)$ is used to convert the analog form to discrete form PID controller. The discrete domain PID controller can be written as:

$$G_c(z) = K_p + \frac{K_i}{1-z^{-1}} + K_d \left(1 - z^{-1}\right) \tag{13.13}$$

The control output of discrete PID controller can be represented as:

$$u[kT_s] = K_p e[kT_s] + u_i\left[(k-1)T\right] + K_i e[kT_s] + K_d\left(e[kT_s] - e\left[(k-1)T_s\right]\right) \tag{13.14}$$

The design constraint for digital PID controller is:

- Low overshoot;
- Less settling time;
- Better frequency response;
- Suitable setpoint regulation, load disturbance rejection and set-point tracking.

13.5   PULSE WIDTH MODULATION (PWM)

Pulse width modulation (PWM) effectively uses a rectangular pulse wave whose pulse width is modulated, which results in variation of average value of the waveform. Figure 13.4 shows the PWM signal.

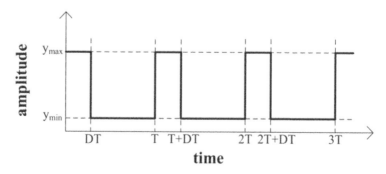

FIGURE 13.4   Pulse width modulated signal.

Let us consider a pulse waveform $f(t)$ with period T, low value y_{min} and a high value y_{max} and a duty cycle D, the average value of waveform is:

$$\bar{y} = \frac{1}{T}\int_0^T f(t)dt \qquad (13.15)$$

As $f(t)$ is a pulse waveform, its value is y_{max} for $0 < t < DT$ and y_{min} for $DT < t < T$

Eqn. (13.15) can be rewritten as:

$$\bar{y} = \frac{1}{T}\left(\int_0^{DT} y_{max}\, dt + \int_{DT}^{T} y_{min}\, dt\right) \qquad (13.16)$$

$$\bar{y} = \frac{1}{T}\left(DTy_{max} + T(1-D)y_{min}\right) \qquad (13.17)$$

$$\bar{y} = \left(Dy_{max} + (1-D)y_{min}\right) \qquad (13.18)$$

For generation of PWM, two signals are required, i.e., original signal also called as modulating signal and carrier signal. The carrier signal can either be a triangle wave or sawtooth wave. The resulted pulse train from both the signals is called modulated signal. The modulated signal is obtained by comparing both the original signal and the carrier signal. Different classification of PWM scheme is provided in Figure 13.5.

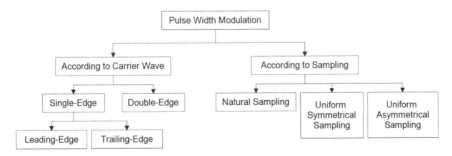

FIGURE 13.5   Classification of pulse width modulated signal.

According to carrier wave, the PWM can be classified as either a single-edge PWM or double-edge PWM. The main difference between single-edge

PWM and double-edge PWM is the type of carrier wave. In single-edge PWM, sawtooth carrier wave is used whereas in double-edge PWM, triangular carrier wave is used. The single-edge PWM can be further classified as leading-edge or trailing-edge. Similarly, according to sampling, PWM can be uniformly sampled or naturally sampled. In naturally sampled PWM, the switching edges of PWM occurs at the same time with the sampling of input signal by the carrier signal. In uniformly sampled PWM (UPWM), the sampling of the input signal occurs at the top or at the bottom of the carrier instead of at the intersection of the input signal and the carrier signal [25].

In single-edge PWM, one of the transition edges is fixed either it is a leading-edge or a trailing-edge while the modulation occurs at the other edge, as the reference signal varies. Depending on the edge, the single-edge PWM is divided in to leading-edge PWM (Figure 13.6(a)) or trailing-edge PWM (Figure 13.6(b)). Double-edge PWM is shown in Figure 13.6(c).

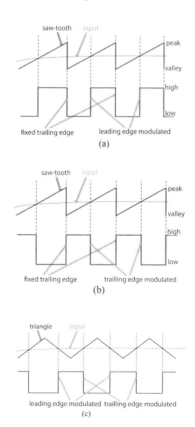

FIGURE 13.6   (a) Leading-edge PWM; (b) trailing edge PWM; and (c) double edge PWM.

Figure 13.7(a) presents leading-edge UPWM and Figure 13.7(b) presents the trailing-edge UPWM. Figure 13.8 presents the naturally sampled PWM.

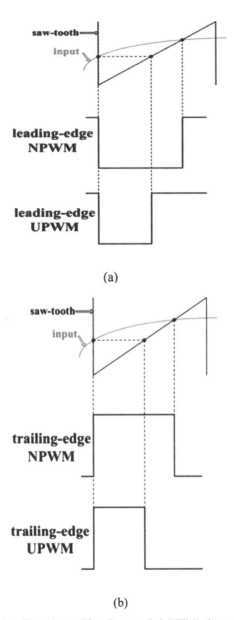

(a)

(b)

FIGURE 13.7   (a) Leading-edge uniformly sampled PWM; (b) trailing edge uniformly sampled PWM.

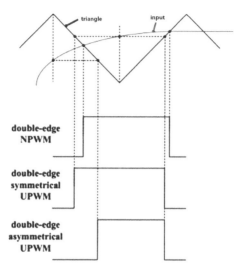

FIGURE 13.8   Naturally sampled PWM.

13.5.1   *ANALOG IMPLEMENTATION OF PULSE WIDTH MODULATION (PWM)*

PWM scheme can either be implemented using multiple operational amplifiers or a dedicated ASIC provides PWM signal of required specification. Figure 13.9 provides the operational amplifier-based implementation of PWM scheme.

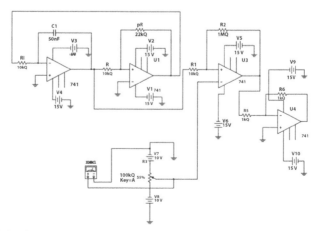

FIGURE 13.9   Operational amplifier based PWM generation.

A review of different ASIC implementation of PWM technique has been provided in Ref. [26]. In 1976, Microsemi Co. developed a monolithic 16 pin DIP ASIC, SG1524 which contains all the control circuitry to regulate a switched-mode power system. After SG1524, different companies manufactured several PWM ICs such as UC1840 from ST Microelectronics, MC3520 from Motorola Inc. Texas Instruments developed UC1524 for PWM regulation and UC1846 to implement fixed frequency current mode control of a switched-mode converter. The primary objective of these ASIC is to reduce the number of components in the control unit. But the op-amp and ASIC based controllers have a significant sensitivity to parameter variation, ambient condition and are not programmable. Figure 13.10 provides a functional block diagram of the PWM generation scheme implemented in ASIC.

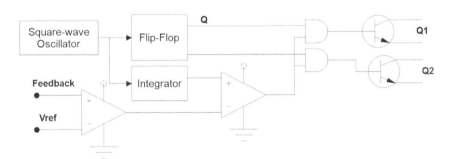

FIGURE 13.10   PWM scheme applied in ASIC.

13.6   DIGITAL PULSE WIDTH MODULATION (DPWM)

There is different type of classical DPWM architecture reported in literature [27]. Table 13.2 provides a comparative performance analysis of different classical DPWM architectures. ASIC and FPGA based implementation of DPWM architecture has been discussed in Ref. [28]. The time-domain analysis of sampling effects in DPWM of DC-DC converter has been discussed in Ref. [29] where relation between sampling delay, duty ratio and multisampling factor has been evaluated:

- Counter based DPWM:
    - o   Presettable down counter and zero detector;
    - o   Cycling counter and comparator;

o   Free running synchronous counter and comparator with data input register.
- Tapped delay-line based DPWM:
  o   Open-loop delay lines;
  o   Closed-loop delay lines;
  o   Feed forward DPWM using open-loop delay lines;
  o   Delay matching network.
- Hybrid DPWM.
- Segmented delay-line DPWM:
  o   Binary weighted delay line DPWM.
- Heterogeneous DPWM.

TABLE 13.2 Comparative Analysis of Classical DPWM

DPWM	Silicon Area	Power	Clock	Delay Element	Clock Freq.
Counter	Lowest	Highest	Yes	0	$2^{nc}f_s$
Tapped delay-line	Highest	Low	No	2^n	fs
Hybrid	Low	High	No	2^{n-nc}	$2^{nc}f_s$
Segmented	High	Lowest	No	2^n	fs
Heterogeneous	Low	Low	No	$2^{nc}f_s$	$2^{nc}f_s$

where n is the total number of delay command; and n_c is the number of bits resolved by counter.

The least significant bit (LSB) of DPWM determines the minimum change in the duty cycle. Thus the resolution of the DPWM is very critical in deciding the accuracy and the overall performance of the converter. Counter-comparator and delay-line DPWM method are unsuitable for high resolution application as it requires higher clock frequency. Dithering and $\sum-\Delta$ PWM [30] improves the resolution but with an additional cost of AC ripple. Hybrid counter delay line operation provides a good trade-off between area, clock frequency and resolution. DPWM can either be of single-phase or multi-phase [31] configuration (Figure 13.11).

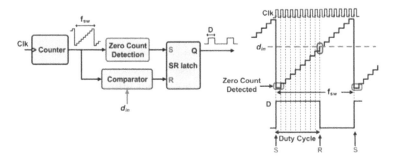

FIGURE 13.11   Counter based DPWM.

13.7   SIMULATION RESULTS

To illustrate the functionality of digital controller in DC-DC converter, the study has considered a DC-DC buck converter. The design specification of the DC-DC buck converter is summarized in Table 13.3.

TABLE 13.3   Circuit Parameters of the Prototype Synchronous Buck Converter

Parameter	Symbol	Nominal Value
Supply voltage	Vi	$12 \pm 10\%$ V DC
Filter inductance	L	2 mH
ESR of inductor	R_L	80 mΩ
Filter capacitor	C	470 µF
ESR of capacitor	r_c	5 mΩ
Maximum load resistance	R_{max}	120 Ω
Minimum load resistance	R_{min}	12 Ω
Switching frequency	f_s	30 kHz
Desired output voltage	V_{ref}	$5 \pm 2\%$ V DC
Nominal load current	I_0	0.41 A
Output voltage ripple	ΔV_0	25 mV (P-P)
Maximum ripple current through inductor	ΔI_L	20% of load current

Substituting the above-mentioned values in Eqn. (13.9), the control to output voltage transfer function of synchronous buck converter is represented by:

$$G(s) = \frac{2.82 \times 10^{-5} s + 12}{1.52 \times 10^{-8} s^2 + 4.585 \times 10^{-5} s + 1}$$ (13.19)

Figure 13.12 shows the controlled output voltage of DC-DC buck converter at R_{max}. Figure 13.13 shows the variation of V_o due to variation in V_i. When the input voltage supplied by a rectifier unit is changed arbitrarily, then the corresponding inductor current also changes significantly.

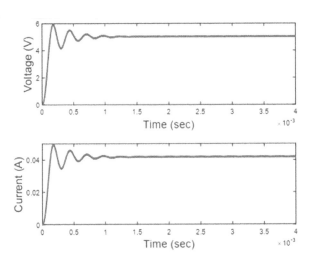

FIGURE 13.12   Controlled output of DC-DC converter at R_{max}.

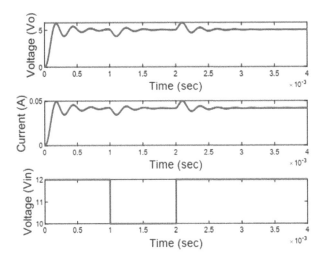

FIGURE 13.13   Variation of V_o of DC-DC converter due to variation in V_i.

13.7.1 EXPERIMENTAL VERIFICATION

For experimental verification of digital controller of DC-DC buck converter, laboratory prototype has been developed. The specification of components for buck converter has been summarized in Table 13.4. For implementation of digital controller module, Artix-7 series FPGA has been used. The resource utilization analysis of FPGA has been summarized in Table 13.5. Figure 13.14 presents the FPGA based implementation of digital control of buck converter. Figure 13.15 presents the experimental input-output of the digital controller of DC-DC buck converter.

TABLE 13.4 Specification of Components for Prototype Development

Device	Part Number	Specification
MOSFET	IRLB8748	$V_{dss} = 30$ V, $R_{DS(ON)} = 4.8$ mΩ
MOSFET driver	IR2110	High and low side driver
Controller	Artix-7 FPGA	–
DPWM bits	10-bit	–

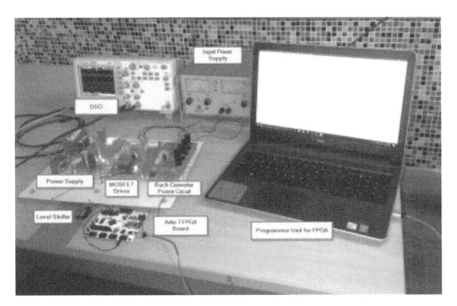

FIGURE 13.14 Developed prototype for digitally controlled buck converter.

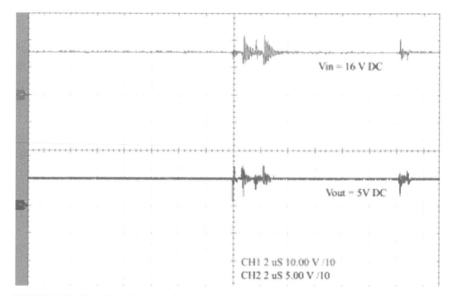

FIGURE 13.15 Experimental input-output waveform for digitally controlled buck converter.

TABLE 13.5 FPGA Resource Utilization Report

Resource	Used	Available	Utilization (%)
Slice LUTs	1,562	20,800	7.51
Slice registers	1,459	41,600	3.51
F7 Muxes	116	16,300	0.71
LUT as logic	1,423	20,800	6.84
LUT as memory	139	9,600	1.45
LUT flip flop pairs	573	20,800	2.75
Block RAM tile	8	50	16.00
DSP	1	90	1.11
Clocking (BUFGCTRL)	3	32	9.38

13.8   CONCLUSION

This study provides a detailed analysis of digital voltage mode control of DC-DC buck converter. Different design aspects of digital controller and digital pulse with modulator scheme have been discussed. Design and mathematical model of buck converter has been derived. Digital PID controller and DPWM have been modeled and closed-loop performance of digital control of DC-DC converter has been presented in this study. For experimental verification, FPGA based implementation of digital controller and DPWM has been presented.

KEYWORDS

- **DC-DC converter**
- **digital pulse width modulation**
- **digitally controlled buck converter**
- **FPGA implementation**
- **PID controller**
- **serial-in serial-out (SISO)**

REFERENCES

1. Buccella, C., Cecati, C., & Latafat, H., (2012). Digital control of power converters a survey. *IEEE Transactions on Industrial Informatics, 8*(3), 437–447.
2. Liu, Y. F., Meyer, E., & Liu, X., (2009). Recent developments in digital control strategies for dc/dc switching power converters. *IEEE Transactions on Power Electronics, 24*(11), 2567–2577.
3. Padhee, S., Pati, U. C., & Mahapatra, K., (2018). Closed-loop parametric identification of dc-dc converter. *Proceedings of the Institution of Mechanical Engineers, Part I: Journal of Systems and Control Engineering, 232*(10), 1429–1438.
4. Duan, Y., & Jin, H., (1999). Digital controller design for switch mode power converters. In: *APEC '99. Fourteenth Annual Applied Power Electronics Conference and Exposition; 1999 Conference Proceedings (Cat. No. 99CH36285)* (Vol. 2, pp. 967–973). IEEE.
5. Oliva, A. R., Ang, S. S., & Bortolotto, G. E., (2006). Digital control of a voltage-mode synchronous buck converter. *IEEE Transactions on Power Electronics, 21*(1), 157–163.
6. Al-Atrash, H., & Batarseh, I., (2007). Digital controller design for a practicing power electronics engineer. In: *APEC 07-Twenty-Second Annual IEEE Applied Power Electronics Conference and Exposition* (pp. 34–41). IEEE.
7. Tu, S. H. L., & Chiu, M. M. H., (2013). Digital pulse-width modulation controller based on fully table look-up for system-on-a-chip applications. *IET Power Electronics, 6*(9), 1778–1785.
8. Corradini, L., Mattavelli, P., Tedeschi, E., & Trevisan, D., (2008). High bandwidth multi-sampled digitally controlled dc-dc converters using ripple compensation. *IEEE Transactions on Industrial Electronics, 55*(4), 1501–1508.

9. Trunti˘c, M., & Milanovi˘c, M., (2014). Voltage and current-mode control for a buck-converter based on measured integral values of voltage and current implemented in FPGA. *IEEE Transactions on Power Electronics, 29*(12), 6686–6699.

10. Sajitha, G., & Kurian, T., (2016). Reduction in limit cycle oscillation and conducted electromagnetic emissions by switching frequency adjustment in digitally controlled dc-dc converters. *EPE Journal, 26*(3), 96–103.

11. Vyas, S., Gupte, A., Gill, C. D., Cytron, R. K., Zambreno, J., & Jones, P. H., (2013). Hardware architectural support for control systems and sensor processing. *ACM Transactions on Embedded Computing Systems (TECS), 13*(2), 16.

12. Liou, W. R., Lacorte, W. B., Caberos, A. B., Yeh, M. L., Lin, J. C., Lin, S. C., & Sun, C. S., (2012). A programmable controller IC for dc/dc converter and power factor correction applications. *IEEE Transactions on Industrial Informatics, 9*(4), 2105–2113.

13. Malinowski, A., & Yu, H., (2011). Comparison of embedded system design for industrial applications. *IEEE Transactions on Industrial Informatics, 7*(2), 244–254.

14. Compton, K., & Hauck, S., (2002). Reconfigurable computing: A survey of systems and software. *ACM Computing Surveys (csuR), 34*(2), 171–210.

15. Fang, Z., Carletta, J. E., & Veillette, R. J., (2005). A methodology for FPGA-based control implementation. *IEEE Transactions on Control Systems Technology, 13*(6), 977–987.

16. Ho, C. H., Yu, C. W., Leong, P., Luk, W., & Wilton, S. J., (2009). Floating-point FPGA: Architecture and modeling. *IEEE Transactions on Very Large Scale Integration (VLSI) Systems, 17*(12), 1709–1718.

17. Kuon, I., & Rose, J., (2007). Measuring the gap between FPGAS and ASICS. *IEEE Transactions on Computer-Aided Design of Integrated Circuits and Systems, 26*(2), 203–215.

18. Fratta, A., Griffero, G., & Nieddu, S., (2004). Comparative analysis among DSP and FPGA-based control capabilities in PWM power converters. In: *30ᵗʰ Annual Conference of IEEE Industrial Electronics Society, 2004* (Vol. 1, pp. 257–262). IEEE.

19. Youness, H., Moness, M., & Khaled, M., (2014). MPSOCS and multicore microcontrollers for embedded PID control: A detailed study. *IEEE Transactions on Industrial Informatics, 10*(4), 2122–2134.

20. Monmasson, E., Idkhajine, L., & Naouar, M. W., (2011). *FPGA-based Controllers* (Vol. 5, No. 1, pp. 14–26). IEEE industrial electronics magazine.

21. Chan, Y. F., Moallem, M., & Wang, W., (2007). Design and implementation of modular FPGA-based PID controllers. *IEEE transactions on Industrial Electronics, 54*(4), 1898–1906.

22. Barai, M., Sengupta, S., & Biswas, J., (2009). Digital controller for DVS enabled dc-dc converter. *IEEE Transactions on Power Electronics, 25*(3), 557–573.

23. Islam, M. M., Allee, D. R., Konasani, S., & Rodr´ıguez, A. A., (2004). A low-cost digital controller for a switching dc converter with improved voltage regulation. *IEEE Power Electronics Letters, 2*(4), 121–124.

24. Kılınc, S., & Cabuk, G., (2015). Spectrum of a frequency-modulated pulse width modulation signal. *International Journal of Circuit Theory and Applications, 43*(3), 390–400.

25. Sahin, O. U., (2017). *The Design of a High Frequency Pulse width Modulation Integrated Circuit with External Synchronization Capability.* Ph.D. dissertation, middle east technical university.

26. Bindra, A., (2014). *Pulse Width Modulated Controller Integrated Circuit: Four Decades of Progress - a Look Back, 1*(3), 10–44. IEEE power electronics magazine.

27. Syed, A., Ahmed, E., Maksimovic, D., & Alarcon, E., (2004). Digital pulse width modulator architectures. In: *2004 IEEE 35th Annual Power Electronics Specialists Conference (IEEE Cat. No. 04CH37551)* (Vol. 6, pp. 4689–4695). IEEE,.

28. Chander, S., Agarwal, P., & Gupta, I., (2013). ASIC and FPGA based DPWM architectures for single-phase and single-output dc-dc converter: A review. *Central European Journal of Engineering, 3*(4), 620–643.

29. Kumar, M., & Gupta, R., (2015). Time-domain analysis of sampling effect in DPWM of dc-dc converters. *IEEE Transactions on Industrial Electronics, 62*(11), 6915–6924.

30. Deblecker, O., (2011). High-resolution DPWM using sigma-delta modulator implemented on a low-cost buck development board. *International Journal of Electrical Engineering Education, 48*(4), 391–404.

31. Foley, R., Kavanagh, R., Marnane, W., & Egan, M., (2006). Multiphase digital pulse width modulator. *IEEE Transactions on Power Electronics, 21*(3), 842–846.

Index